A Student's Guide to Dimensional Analysis

This introduction to dimensional analysis covers the methods, history, and formalization of the field, and provides physics and engineering applications. Covering topics from mechanics, hydro- and electrodynamics to thermal and quantum physics, it illustrates the possibilities and limitations of dimensional analysis. Introducing basic physics and fluid engineering topics through the mathematical methods of dimensional analysis, this book is perfect for students in physics, engineering, and mathematics. Explaining potentially unfamiliar concepts such as viscosity and diffusivity, this text includes worked examples and end-of-chapter problems with answers provided in an accompanying appendix, which help make this text ideal for self-study. Long-standing methodological problems arising in popular presentations of dimensional analysis are also identified and solved, making it a useful text for advanced students and professionals.

DON S. LEMONS is Professor Emeritus of Physics at Bethel College in North Newton, Kansas, and has served as an assistant editor of the *American Journal of Physics*. He is a member of the American Physical Society and his research focuses primarily on plasma physics.

A Student's Guide to Dimensional Analysis

DON S. LEMONS
Bethel College, Kansas

CAMBRIDGE
UNIVERSITY PRESS

CAMBRIDGE
UNIVERSITY PRESS

University Printing House, Cambridge CB2 8BS, United Kingdom

One Liberty Plaza, 20th Floor, New York, NY 10006, USA

477 Williamstown Road, Port Melbourne, VIC 3207, Australia

4843/24, 2nd Floor, Ansari Road, Daryaganj, Delhi – 110002, India

79 Anson Road, #06–04/06, Singapore 079906

Cambridge University Press is part of the University of Cambridge.

It furthers the University's mission by disseminating knowledge in the pursuit of education, learning, and research at the highest international levels of excellence.

www.cambridge.org
Information on this title: www.cambridge.org/9781107161153
10.1017/9781316676165

First published 2017

Printed in the United Kingdom by TJ International Ltd. Padstow Cornwall

A catalogue record for this publication is available from the British Library.

Library of Congress Cataloging-in-Publication Data
Names: Lemons, Don S. (Don Stephen), 1949– author.
Title: A student's guide to dimensional analysis / Don Lemons.
Description: New York : Cambridge University Press, 2017. | Includes bibliographical references and index.
Identifiers: LCCN 2016040385 | ISBN 9781107161153 (hardback)
Subjects: LCSH: Dimensional analysis. | BISAC: SCIENCE / Mathematical Physics.
Classification: LCC TA347.D5 L46 2017 | DDC 530.8–dc23 LC record available at https://lccn.loc.gov/2016040385

ISBN 978-1-107-16115-3 Hardback
ISBN 978-1-316-61381-8 Paperback

Contents

Preface

I was charmed when as a young student I watched one of my physics professors, the late Harold Daw, work a problem with dimensional analysis. The result appeared as if by magic without the effort of constructing a model, solving a differential equation, or applying boundary conditions. But the inspiration of the moment did not, until many years later, bear fruit. In the meantime my acquaintance with this important tool remained partial and superficial. Dimensional analysis seemed to promise more than it could deliver.

Dimensional analysis has charmed and disappointed others as well. Yet there is no doubt that a deep understanding of its methods is useful to researchers in a number of fields. Attesting to this fact is that dimensional analysis is the subject of several good, graduate-level monographs. Even so, dimensional analysis is often ignored at the introductory level except when teachers admonish their students to "check the units of their result" and warn them against "adding apples and oranges."

The problem for teachers and students is that dimensional analysis has no settled place in the physics curriculum. The mathematics required for its application is quite elementary – of the kind one learns in a good high school course – and its foundational principle is essentially a more precise version of the rule against "adding apples and oranges." Yet the successful application of dimensional analysis requires physical intuition – an intuition that develops only slowly with the experience of modeling and manipulating physical variables. But how much intuition is required?

I have written *A Student's Guide to Dimensional Analysis* in the belief that the simple techniques of dimensional analysis can deepen our understanding and enhance our exploration of physical situations and processes at the introductory level. Thus, this text is designed for students who are taking or have taken an entry-level, mathematically oriented, university physics course. More experienced students and professionals may also find this text useful.

One elementary application of dimensional analysis is to a simple pendulum, that is, to a compact weight swinging from the end of a string. How long does the pendulum take to complete one cycle of its motion? It seems that this interval Δt might depend on the mass of the weight m, the acceleration of gravity g, the length of the string l, and the maximum angle of its swing θ. The adjective "simple" in "simple pendulum" means that the size of the weight is negligibly small compared to the length of the pendulum l. Therefore, we search for a function of m, l, g, and θ that produces Δt. We note that the International System or SI unit of the period Δt is the second, the SI unit of mass m is the kilogram, of length l is the meter, and of acceleration g is a meter divided by a second squared. The angle θ is dimensionless. We quickly discover that $\sqrt{l/g}$ is the only combination of m, l, and g that produces a quantity whose SI unit is a second. Therefore, it must be that

$$\Delta t = \sqrt{\frac{l}{g}} \cdot f(\theta)$$

where $f(\theta)$ is an as yet undetermined, dimensionless function of θ.

This result is typical of dimensional analysis. Much is learned but something is left unlearned. The analysis teaches us that the period Δt of a simple pendulum is directly proportional to the square root of its length $\sqrt{l}$ and does not at all depend on its mass m. We could experimentally verify these findings, but dimensional analysis suggests that limited time might be better spent exploring the undetermined function $f(\theta)$. However, this example misleads as well as informs. Not many applications of dimensional analysis are this simple, nor are we usually aware, as we probably are in this case, of the result beforehand.

Dimensional analysis is most revealing when we know *something* but not *everything* about a situation or process. Then dimensional analysis builds on that *something* we already know. We might, for instance, be familiar with the equations that describe a certain process but not have the skill or the time to solve them in the usual way. Or we might wish to expand our preliminary knowledge of a solution before completely determining its form. Then again, we might know only the category of the problem (that is, mechanics, thermodynamics, or electrodynamics) and the variable we want to determine (that is, an oscillation period, a pressure drop, or an energy loss rate). In all these cases, dimensional analysis constrains how the relevant physical variables and constants work together to produce the result sought.

The techniques of dimensional analysis could be presented in one well-chosen example, occupying one or two pages of text. Then one might, with some reason, expect to apply these techniques to new problems as they arise. But this expectation is bound to remain unrealized. In order for dimensional analysis to be fruitful, it must first be cultivated. Exploring the method's motivation, its history, and its formalization are steps in this cultivation. Examining simple applications is another. Such are the aims of Chapter 1 "Introduction."

Chapters 2 through 6 take up more complex but still introductory examples of dimensional analysis grouped into several subject areas: Mechanics, Hydrodynamics, Thermal Physics, Electrodynamics, and Quantum Physics. These examples illustrate the possibilities and limitations of dimensional analysis. As appropriate, I will explain concepts (such as surface tension, viscosity, and diffusivity) that may not be familiar to all readers.

In the final chapter, Chapter 7 "Dimensional Cosmology," I use dimensional analysis to take a few steps in the direction of uncovering the dimensional structure of our world. The result, preliminary and partial as it is, brings us close to the boundary separating the known from the unknown.

Dimensional analysis has no settled place in the physics curriculum because it fits easily in any number of places. We can use it in elementary fashion to recall the shape of a formula or to reaffirm our understanding of an already solved problem. Or we can use it to push forward into territory unknown to us or uncharted by anyone.

Sometimes dimensional analysis fails us, but it is unlikely to do so without announcing its failure and suggesting a better way to proceed. Does the method produce no result? Then we have left out a crucial variable or constant. Is the result uninformative? Then we have included too many variables and constants.

A Student's Guide to Dimensional Analysis is designed to guide readers to an understanding of the motivation, methods, and exemplary applications of dimensional analysis, its scope and powers as well as its limitations. But I also hope the text will recreate for you the charm and magic that first attracted me to the subject years ago.

Acknowledgments

One of the rewards of writing a book is receiving responses to its initial drafts from an obliging set of friends. I have been very fortunate in my "first responders." Ralph Baerlein and Galen Gisler read the whole text, suggested many improvements, and kept me from making a number of mistakes. Ikram Ahmed, Christopher Earles, Rick Shanahan, and David Watkins all reviewed parts of the text and gave me good advice. Jeff Martin suggested the problem "cooking a turkey." I offer a heartfelt "thanks" to them all.

My participation in a series of seminars on similarity methods, organized by Susan G. Sterrett of Wichita State University in the academic year 2014–15, renewed a youthful interest in dimensional analysis. As usual, Cambridge University Press has provided me with excellent editorial services.

1

Introduction

1.1 Dimensional Homogeneity

Most physical variables and constants have dimensions. A mass, a distance, and a time have, respectively, the dimensions *mass*, *length*, and *time*. Dimensions are often taken for granted. They slip beneath our notice. Yet dimensions are important elements of the way we think about the physical world.

Imagine an animal that grows in size while keeping roughly the same shape. Galileo (1584–1642) reasoned that the weight of the animal increases with its volume in direct proportion to the third power l^3 of a characteristic length l, say, for an elephant, the length of its foreleg. Because animal limbs push and pull across a cross-section, the strength of an animal increases with the cross-sectional area of its limbs, that is, in direct proportion to l^2. Thus the animal's strength-to-weight ratio changes in proportion as l^2/l^3, that is, as $1/l$ or l^{-1}. Therefore, larger animals are less able to support their weight than smaller ones are. Galileo illustrated this conclusion by comparing the relative strength of dogs and horses – creatures with roughly the same shape.

> A small dog could probably carry on his back two or three dogs of his own size; but I believe that a horse could not carry even one of his own size. [1]

If Galileo had not thought dimensionally, he could not have made this interesting argument.

By the time of Isaac Newton (1643–1727) scientists had begun to think in terms of combinations of different dimensions. For instance, the dimension of speed is *length* divided by *time*, the dimension of acceleration is *length* divided by *time squared*, and, according to Newton's second law, the dimension of force is *mass* times *length* divided by *time squared*. Newton regarded *mass*, *length*, and *time* as primary, fundamental dimensions and combinations of these as secondary, derived ones. [2]

1

One of the first things a physics student learns is that one should not add, subtract, equate, or compare quantities with different dimensions or quantities with the same dimension and different units of measure. For instance, one cannot add a mass to a length or, for that matter, 5 meters to 2 kilometers. This rule against what is sometimes called "adding apples and oranges" means that every term that is added, subtracted, equated, or compared in every valid equation or inequality must be of the same dimension denominated in the same unit of measure. This is the *principle of dimensional homogeneity*.

The principle of dimensional homogeneity is nothing new. Scientists have long assumed that every term in every fully articulated equation that accurately describes a physical state or process has the same dimension denominated in the same unit of measure. However, it was not until 1822 that Joseph Fourier (1768–1830) expressed this principle in a way that allowed important consequences to be derived from it. [3]

Symmetry under Change of Units

Behind the principle of dimensional homogeneity is a symmetry principle. Symmetry principles tell us that something remains the same as something else is changed. Here the *something that remains the same* is the form of the equation or inequality and the *things that are changed* are the units in which the dimensions of its terms are expressed. Thus, if we change the unit of length from meters to kilometers and the form of the equation does not change, that equation observes this particular symmetry and is, at least in this regard, dimensionally homogeneous. If we change all of its units and the form of the equation does not change, this equation is fully dimensionally homogeneous.

An equation can be useful without being dimensionally homogeneous. For instance,

$$s = 4.9t^2 \tag{1.1}$$

correctly describes the downward displacement s denominated in meters of an object falling freely from rest for a period of time t denominated in seconds. Yet (1.1) is not invariant with respect to changes in its units of measure. Compare (1.1) with

$$s = \frac{1}{2}gt^2 \tag{1.2}$$

in which we have parameterized the acceleration of gravity with the symbol g. This equation is now symmetric with respect to changes in all its units of measure. It is fully articulated and dimensionally homogeneous.

We are concerned in this text with relations among dimensional variables, for instance, s and t, and dimensional constants, such as g, that are symmetric with respect to changes in units and, therefore, dimensionally homogeneous. The principle of dimensional homogeneity and its consequences are foundational to the theory of dimensional analysis.

1.2 Dimensionless Products

Consider the vertical position y of a freely falling object at time t. We know that

$$y - y_o = v_{yo}t - \frac{gt^2}{2} \tag{1.3}$$

where $y - y_o$ is the object's displacement from its initial position y_o, v_{yo} is its initial velocity, g is the magnitude of the gravitational acceleration, and our coordinate system is oriented so that y becomes more negative as the object falls and time advances. Equation (1.3) observes the principle of dimensional homogeneity. For the dimension of $y - y_o$ is *length*; the dimensions of v_{yo} and t are, respectively, *length/time* and *time* so that the dimension of their product $v_{yo}t$ is *length*; and the dimension of gt^2 is *length/time²* multiplied by *time²* or, again, *length*. Furthermore, (1.3) contains no dimensional constants masquerading as dimensionless numbers – as does (1.1).

One consequence of the dimensional homogeneity of (1.3) is that dividing each of its terms by gt^2 produces an equation,

$$\frac{y - y_o}{gt^2} = \frac{v_{yo}}{gt} - \frac{1}{2}, \tag{1.4}$$

that relates one dimensionless combination or "product" $(y - y_o)/gt^2$ to another v_{yo}/gt. This transformation of a dimensionally homogeneous equation, from a relation among dimensional variables and constants to a relation among dimensionless products, can always be realized.

Consider, for instance, the Stefan-Boltzmann law, according to which the density of radiant energy E in a cavity of volume V whose walls are at a temperature T is described by

$$\frac{E}{V} = \frac{8\pi^5}{15} \frac{k_B^4 T^4}{c^3 h^3} \tag{1.5}$$

where k_B is Boltzmann's constant, c is the speed of light, and h is Planck's constant. Each of the variables E, V, and T and each of the constants k_B, c, and

h are dimensional quantities. If equation (1.5) is dimensionally homogeneous (and it is), it may take the form

$$\frac{Ec^3h^3}{Vk_B^4T^4} = C \qquad (1.6)$$

of a dimensionless product $Ec^3h^3/\left(Vk_B^4T^4\right)$ equal to a dimensionless number C. In this case $C = 8\pi^5/15$.

Dimensional Analysis

We have, in these two examples, turned dimensionally homogeneous relations among dimensional variables and constants, (1.3) and (1.5), into relations among one or more dimensionless products, (1.4) and (1.6). We shall soon learn a way to reverse this process. We will first use an algorithm, the Rayleigh algorithm, to discover the dimensionless products relevant to a particular state or process. When only one dimensionless product is found, the only way it can form a dimensionally homogeneous equation is for this product to equal some dimensionless number as in (1.6). When two or more dimensionless products are found, as in (1.3), they must be related to one another by some function, as in (1.4). The Rayleigh algorithm does not determine these numbers and these functions but merely finds the dimensionless products.

1.3 Dimensional Formulae

Every dimensional variable or constant assumes values in the form of a number times a unit of measure – for instance, 5 kilograms or 16 meters. Furthermore, every unit of measure makes its dimension known. A meter per second and a kilometer per hour are both a *length/time* while a metric ton and a kilogram are both *masses*. We need to know the dimension, more precisely the dimensional formula, of every relevant dimensional variable and constant in order to dimensionally analyze a state or process. For this purpose we use the symbol M to stand for the dimension *mass*, L to stand for the dimension *length*, and T to stand for the dimension *time*. The notation $[x]$ means "the dimension of x." Therefore, $[m] = M$ and $[g] = LT^{-2}$ are dimensional formulae. Not every dimensional formula can be expressed in terms of only M, L, and T, but many can be.

The dimensional formula of a product of factors is the product of the dimensional formula of each factor. Thus $[ma] = [m][a]$. For convenience, we define the dimensional formula of a dimensionless number to be 1. Therefore, $[\pi] = 1$ and so $[9.8 \cdot m/s^2] = [9.8] \cdot [m/s^2] = [m] \cdot [s^{-2}] = LT^{-2}$.

1.4 The Rayleigh Algorithm

John William Strutt (1842–1919), also known as Lord Rayleigh, successfully applied dimensional analysis to a number of problems over a long career. He dimensionally analyzed the strength of bridges, the velocity of waves on the surface of water, the vibration of tuning forks and drops of falling water, the color of the sky, the decay of charge on an electrical circuit, the determinants of viscosity, and the flow of heat from a hot object immersed in a cool stream of water. Rayleigh prefaced his 1915 summary of these applications of the principle of dimensional homogeneity (known to him as "the principle of similitude") with these words,

> I have often been impressed with the scanty attention paid even by original workers to the great principle of similitude. It happens not infrequently that results in the form of "laws" are put forth as novelties on the basis of elaborate experiments, which might have been predicted *a priori* after a few minutes' consideration. [4]

While our applications of dimensional analysis may require more than "a few minutes consideration," Rayleigh's method of applying "the principle of similitude" is simple and direct. We adopt it, as have many others, with only slight modification.

A Marble on the Interior Surface of a Cone

To illustrate Rayleigh's method, imagine a small marble of mass m rolling in a circle of radius R on the interior surface of an inverted cone defined by an angle θ as illustrated in Figure 1.1. We wish to know how the time Δt required for the marble to complete one orbit is determined by m, R, and θ. The acceleration of gravity g may also enter into the relation we seek. Gravity,

Figure 1.1. Marble on the interior surface of a cone defined by angle θ.

after all, is one of the two forces that keep the marble on the cone's surface. The intermolecular forces of the material composing the cone and the marble also determine, in some degree, the period Δt, but we ignore these forces because we believe their effect is adequately accounted for by assuming the marble stays on the surface of the cone. By including some variables and constants in our analysis and excluding others we construct a model of the marble's motion.

Rayleigh's Algorithm

Rayleigh's method of dimensional analysis identifies the dimensionless products one can form out of the model variables and constants, in this case Δt, m, R, g, and θ. Each dimensionless product takes a form $\Delta t^\alpha m^\beta R^\gamma g^\delta \theta^\varepsilon$ determined by the Greek letter exponents α, β, γ, δ, and ε or, somewhat more simply, by the form $\Delta t^\alpha m^\beta R^\gamma g^\delta$ and the exponents α, β, γ, and δ. After all, the angle θ, whether denominated in radians or degrees, is proportional to a ratio of an arc length to a radius, that is, a ratio of one length to another. While angles have units (degrees or radians), their units are dimensionless.

The key to Rayleigh's method of finding dimensionless products is to require that the product $\Delta t^\alpha m^\beta R^\gamma g^\delta$ be dimensionless. Since

$$
\begin{aligned}
\left[\Delta t^\alpha m^\beta R^\gamma g^\delta\right] &= \left[\Delta t^\alpha\right]\left[m^\beta\right]\left[R^\gamma\right]\left[g^\delta\right] \\
&= [\Delta t]^\alpha [m]^\beta [R]^\gamma [g]^\delta \\
&= T^\alpha M^\beta L^\gamma \left(LT^{-2}\right)^\delta \\
&= T^{\alpha-2\delta} M^\beta L^{\gamma+\delta},
\end{aligned}
\tag{1.7}
$$

the product $\Delta t^\alpha m^\beta R^\gamma g^\delta$ is dimensionless when

$$
T : \alpha - 2\delta = 0,
\tag{1.8a}
$$

$$
M : \beta = 0,
\tag{1.8b}
$$

and

$$
L : \gamma + \delta = 0.
\tag{1.8c}
$$

The three equations (1.8a), (1.8b), and (1.8c) constrain the four unknowns, α, β, γ, and δ, to a family of solutions $\beta = 0$, $\gamma = -\alpha/2$, and $\delta = \alpha/2$ parameterized by α. [The symbols T, M, and L preceding equations (1.8) identify the source of each constraint.] Therefore, $\left(\Delta t g^{1/2}/R^{1/2}\right)^\alpha$ is dimensionless for any α, which means that $\Delta t g^{1/2}/R^{1/2}$, as well as θ, is dimensionless.

Once we know the dimensionless products that can be formed out of the model's dimensional variables and constants, we know they must be related to one another by an undetermined function, that is, in this case expressed by

$$\Delta t = \sqrt{\frac{R}{g}} \cdot f(\theta) \tag{1.9}$$

where $f(\theta)$ is a dimensionless function of the dimensionless "product" θ. This is as far as dimensional analysis *per se* takes us. A more detailed, dynamical study reveals that $f(\theta) = 2\pi\sqrt{\tan\theta}$.

The Rayleigh Algorithm Modified

Note that equations (1.8) are solved by a family of solutions parameterized by a non-vanishing exponent α and also that the identity of the dimensionless product $\Delta t g^{1/2}/R^{1/2}$ is independent of the value of α. The exponent α, introduced in the term Δt^{α}, seems superfluous, and indeed, it is – as long as we know that Δt, the variable whose expression we seek, the *variable of interest*, remains in the dimensionless product. In this case no harm is done by freely choosing α. In particular, choosing $\alpha = 1$ is equivalent to determining the three remaining exponents β, γ, and δ as those that make $\Delta t m^{\beta}R^{\gamma}g^{\delta}$ dimensionless. Then $\beta = 0$, $\gamma = -1/2$, and $\delta = 1/2$. This solution again produces the dimensionless product $\Delta t g^{1/2}/R^{1/2}$. Henceforth we adopt the practice of including the variable of interest with an exponent of 1 as the first factor in the dimensionless product.

Observe that this analysis, issuing as it does in $\Delta t g^{1/2}/R^{1/2} = f(\theta)$, is a significant advance on knowing only that Δt, m, R, g, and θ are related to one another by an unknown function, say, by $\Delta t = h(m, R, g, \theta)$. For suppose that empirically determining the function $f(\theta)$ in (1.9) requires 10 pairs of $\Delta t g^{1/2}/R^{1/2}$ versus θ data. Since 10 pairs of data determine how one term depends on one other (the others remaining constant), 10^4 pairs of data are required to determine how one variable Δt depends on the four dimensional variables and constants m, R, g, and θ. Thus, 10^4 pairs of data are required to determine the function in $\Delta t = h(m, R, g, \theta)$. The Rayleigh algorithm reduces the effort required by a factor of 1,000!

1.5 The Buckingham π Theorem

In 1914, Edgar Buckingham (1867–1940) proved, in formal algebraic detail, a theorem we have, thus far, merely illustrated – a theorem usually referred to as the *Buckingham π theorem* or sometimes, more simply, as the *π theorem*. [5] The π theorem may be divided into two conceptually distinct parts. First,

> If an equation is dimensionally homogeneous, it can be reduced to a relationship
> among a complete set of independent dimensionless products. [6]

A set of dimensionless products is *complete* if and only if all possible dimensionless products of the dimensional variables and constants can be expressed as a product of powers of the members of this set. The members of this set are *independent* if and only if none of them can be expressed as a product of powers of the other members.[a] The symbol π in the phrase π *theorem* refers to members of a complete set of independent dimensionless products. Buckingham denoted these dimensionless products by $\pi_1, \pi_2 \ldots$ Thus, for example, in the marble on the interior of a cone problem, $\pi_1 = \Delta t g^{1/2}/R^{1/2}$ and $\pi_2 = \theta$. The second part of the π theorem consists of the following statement.

> The number of complete and independent dimensionless products N_p is equal to
> the number of dimensional variables and constants N_V that describe the state or
> process minus the minimum number of dimensions N_D needed to express their
> dimensional formulae. Thus,

$$N_p = N_V - N_D. \tag{1.10}$$

Statement (1.10) is the most common expression of the π theorem.

1.6 The Number of Dimensions

Most dimensional analysts adopt M, L, and T as dimensions appropriate for mechanical processes and states. We did so in describing the marble on the interior surface of a cone. In that case, $N_D = 3$. Furthermore, since Δt, m, R, g, and θ describe the marble's motion, $N_V = 5$. Therefore, according to (1.10) $N_P = N_V - N_D$, $2(= 5 - 3)$ complete and independent dimensionless products should be produced. By applying the Rayleigh algorithm we find these to be $\Delta t g^{1/2}/R^{1/2}$ and θ. The set $\Delta t g^{1/2}/R^{1/2}$ and θ is complete because every possible dimensionless product of Δt, m, R, g, and θ can be expressed as a product of some power of $\Delta t^2 R/g$ times some power of θ. And its members are independent because $\Delta t^2 R/g$ and θ are not powers of each other.

But the *minimum* number of dimensions needed to express the dimensional formulae of the N_V dimensional variables and constants is not always 3, as it is in this example. Neither is the identity of the minimum number of dimensions necessarily M, L, and T – as they often are in mechanical problems. Rather, the dimensions required are, in Buckingham's words, the "arbitrary fundamental

---

[a] A complete set of independent products plus all dimensionless products that can be formed from them is itself a *group* because these products: (a) are closed under multiplication, (b) contain an identity element 1, and (c) each product π_i has an inverse π_i^{-1}.

units [dimensions] needed as a basis for the absolute system." [7] And only when we can ensure that N_D is the *minimum* number of dimensions needed can we depend on $N_P = N_V - N_D$ to be observed. Otherwise, $N_P = N_V - N_D$ remains a mere "rule of thumb" – often observed but sometimes not. We will learn how to recognize the minimum number of dimensions in Section 2.2.

1.7 The Number of Dimensionless Products

Note that the more dimensionless products, $\pi_1, \pi_2, \ldots \pi_{N_P}$, produced, the less determined the state or process described by the dimensional model. After all, if only one product π_1 is produced, the result sought assumes a form $f(\pi_1) = 0$ whose solution $\pi_1 = C$ is in terms of a single undetermined dimensionless number C. However, if two dimensionless products, π_1 and π_2, are produced, these are related by $g(\pi_1, \pi_2) = 0$ whose solution $\pi_1 = h(\pi_2)$ leaves a function $h(\pi_2)$ of a single variable undetermined. And if three dimensionless products, π_1, π_2, and π_3, are produced, these are related by a function $j(\pi_1, \pi_2, \pi_3) = 0$ whose solution $\pi_1 = k(\pi_2, \pi_3)$ leaves a function $k(\pi_1, \pi_2)$ of two variables undetermined.

It is clear that in order to more completely determine a state or process, we need to minimize the number N_P of complete and independent dimensionless products. According to the rule of thumb $N_P = N_V - N_D$, we do this by minimizing N_V (the number of dimensional variables and constants that describe the model) and, assuming we have such freedom, by maximizing N_D (the minimum number of dimensions in terms of which these variables and constants can be expressed.) However, minimizing N_V and maximizing N_D are not straightforward tasks. Both require skill and judgment – the same kind of skill and judgment needed to construct a model of a physical state or process.

1.8 Example: Pressure of an Ideal Gas

Many of these ideas are illustrated in the dimensional analysis of how the pressure p of an ideal gas depends on quantities that describe its state. The pressure a gas exerts on its container walls is the average rate per unit area with which its molecules collide with and transfer momentum to the wall. The ideal gas model treats these molecules as randomly and freely moving, massive, point particles whose instantaneous collisions with other particles and with the walls conserve their energy.

Table 1.1

Symbol	Description	Dimensional Formula
p	Pressure	$ML^{-1}T^{-2}$
N/V	Number density	L^{-3}
m	Molecular mass	M
$\bar{v}$	Characteristic speed	LT^{-1}

Therefore, we believe the ideal gas pressure p should depend on the number density of the gas molecules N/V where N is the number of gas molecules contained in volume V, the mass of each of the molecules m, and their average or characteristic speed $\bar{v}$. These parameters should be sufficient, since they are the elements out of which the momentum of the gas particles, the rate at which they collide with the wall, and their energy are composed. To include other variables or constants such as, for instance, the acceleration of gravity g would be to introduce extraneous dimensionless products and make our result not so much inaccurate as uninformative. For convenient reference, we collect these symbols, their descriptions, and their dimensional formulae in Table 1.1.

Note that we have included the number of particles N in volume V only in the combination N/V. For this reason, we have $4(= N_V)$ variables: $p, N/V, m$, and $\bar{v}$. Since they are expressed in terms of $3(= N_D)$ dimensions, M, L, and T, the rule of thumb $N_P = N_V - N_d$ predicts 1 $(= 4 - 3)$ dimensionless product.

Recall that in executing the Rayleigh algorithm, here and elsewhere, we enter the variable of interest, the one whose expression in terms of other variables we seek, in this case the gas pressure p, with an exponent of 1 in the first position of the product $p(N/V)^{\alpha}m^{\beta}\bar{v}^{\gamma}$. Then we find the three exponents, α, β, and γ, that render this product dimensionless. Thus,

$$\left[p(N/V)^{\alpha}m^{\beta}\bar{v}^{\gamma}\right] = \left(ML^{-1}T^{-2}\right)\left(L^{-3}\right)^{\alpha}M^{\beta}\left(LT^{-1}\right)^{\gamma}$$
$$= M^{1+\beta}L^{-1-3\alpha+\gamma}T^{-2-\gamma} \qquad (1.11)$$

and, therefore, the exponents must be solutions of

$$M : 1 + \beta = 0, \qquad (1.12a)$$

$$L : -1 - 3\alpha + \gamma = 0, \qquad (1.12b)$$

and

$$T : -2 - \gamma = 0. \qquad (1.12c)$$

From equations (1.12) we find that $\alpha = -1$, $\beta = -1$, and $\gamma = -2$. Therefore, the dimensionless product is $p(V/N)\left(m\bar{v}^2\right)^{-1}$, and so

$$pV = C \cdot Nm\bar{v}^2 \qquad (1.13)$$

where C is an undetermined dimensionless number.

While interesting, this result is not quite what is known as *Boyle's law*. The latter claims that the pressure p of a gas in thermal equilibrium with a constant-temperature environment is inversely proportional to its volume V, that is, $p \propto 1/V$. Equation (1.13) is equivalent to Boyle's law only if $m\bar{v}^2$ is in one-to-one correspondence with the temperature – which, in the ideal gas model, it is.

If we had allowed the volume V to enter the calculation independently of the number density N/V, then the number of variables, p, N, V, m, and $\bar{v}$, would increase to 5 $(N_V = 5)$ while the number of dimensions, M, L, and T, would remain at 3 $(N_D = 3)$. According to the rule of thumb $N_P = N_V - N_D$, the analysis will produce $2(= N_P)$ independent, dimensionless products. Since $PV/\left(m\bar{v}^2\right)$ and N are two such independent dimensionless "products," we need look no further. The result is $pV/m\bar{v}^2 = f(N)$ where $f(N)$ is an undetermined function – a much less informative result than is $pV/Nm\bar{v}^2 = C$. In general, if our model assumptions allow us to reduce the number of dimensional variables and constants, we should do so.

1.9 A Mistake to Avoid

A model of a state or process incorporates certain idealizations and simplifications. Skill and judgment are required to decide which quantities are needed to describe the state or process and what idealizations and simplifications should be incorporated. Similar skill and judgment are required in dimensional analysis, for the *analysis* in dimensional *analysis* is the analysis of a *model*. And the model we adopt in a dimensional analysis is determined by the dimensional variables and constants we adopt and the dimensions in terms of which they are expressed.

Does the object change shape or does it remain rigid? Does the fluid become turbulent or does it remain laminar? Does the hot object radiate or does it merely conduct? While a certain part of dimensional analysis reduces to the algorithmic, no algorithm helps us answer these questions. Rather, our answers define the state or process we describe and the model we adopt.

We will, on occasion, make mistakes. And the mistakes we make in adopting and dimensionally analyzing a model are of several kinds. The best way to identify these mistakes is to identify them in particular examples.

Table 1.2

ω	Frequency	T^{-1}
m	Mass	M
k	Spring constant	MT^{-2}
ρ	Mass density	ML^{-3}
V	Volume	L^3
g	Acceleration of gravity	LT^{-2}

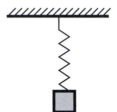

Figure 1.2. Spring-mass system.

Suppose, for instance, I am a relatively inexperienced modeler. And I try to find an expression for the frequency of oscillation ω of a mass m of an extended body of density ρ and volume V that hangs from the end of a spring characterized by spring constant k – as illustrated in Figure 1.2. Since the weight of the mass pulls on the spring, I judge that the acceleration of gravity g is also relevant. The symbols, descriptions, and dimensional formulae in Table 1.2 summarize this, rather naïve, model.

Since there are $6(=N_V)$ dimensional variables and constants and $3(=N_D)$ dimensions, the dimensional analysis should, according to the rule of thumb $N_P = N_V - N_D$, produce $3(=N_P)$ dimensionless products. Accordingly, we seek exponents α, β, γ, δ, and ε that make the product $\omega m^\alpha k^\beta \rho^\gamma V^\delta g^\varepsilon$ dimensionless. We find that

$$
\begin{aligned}
\left[\omega m^\alpha k^\beta \rho^\gamma V^\delta g^\varepsilon\right] &= T^{-1} M^\alpha \left(MT^{-2}\right)^\beta \left(ML^{-3}\right)^\gamma \left(L^3\right)^\delta \left(LT^{-2}\right)^\varepsilon \\
&= T^{-1-2\beta-2\varepsilon} M^{\alpha+\beta+\gamma} L^{-3\gamma+3\delta+\varepsilon},
\end{aligned} \tag{1.14}
$$

and, therefore, that the exponents are constrained by

$$ T : -1 - 2\beta - 2\varepsilon = 0, \tag{1.15a} $$

$$ M : \alpha + \beta + \gamma = 0, \tag{1.15b} $$

and

$$ L : -3\gamma + 3\delta + \varepsilon = 0. \tag{1.15c} $$

Equations (1.15) are solved by $\alpha = 1/2 - \delta + 4\varepsilon/3$, $\beta = -1/2 - \varepsilon$, and $\gamma = \delta - \varepsilon/3$. Consequently,

$$\omega m^\alpha k^\beta \rho^\gamma V^\delta g^\varepsilon = \left(\frac{\omega m^{1/2}}{k^{1/2}}\right)\left(\frac{\rho V}{m}\right)^\delta \left(\frac{m^{4/3}g}{k\rho^{1/3}}\right)^\varepsilon \tag{1.16}$$

for arbitrary δ and ε. As expected the analysis produces three dimensionless products: $\omega m^{1/2}/k^{1/2}$, $\rho V/m$, and $m^{4/3}g/(k\rho^{1/3})$. The result

$$\omega = \sqrt{\frac{k}{m}} \cdot f\left(\frac{\rho V}{m}, \frac{m^{4/3}g}{k\rho^{1/3}}\right), \tag{1.17}$$

where $f(x,y)$, is an undetermined function of two variables, formally expresses the frequency of oscillation ω in terms of the other dimensional variables and constants. Even so, (1.17) by itself tells us little or nothing. We expected something more informative. What went wrong?

The Mistake of Not Minimizing the Number of Dimensional Variables and Constants

One mistake has been to ignore a relation among the variables and constants implied by the model. Clearly, the mass m of the object on the end of the spring is related to its (presumed constant) mass density ρ and its volume V by $m = \rho V$. If so, then (1.17) reduces to

$$\omega = \sqrt{\frac{k}{m}} f\left(1, \frac{m^{4/3}g}{k\rho^{1/3}}\right) \tag{1.18}$$

where $f(1,x)$ is now an undetermined function of only one variable.

Equation (1.18) also follows from recognizing that since $m = \rho V$, the set of model variables and constants need include only two of the following three variables: m, ρ, and V. This reduces N_V from 6 to 5. Since the number of dimensions N_D remains the same at 3, the number of dimensionless products $N_P(= N_V - N_D)$ shrinks to $2(= 5 - 3)$: $\omega m^{1/2}/k^{1/2}$ and $m^{4/3}g/(k\rho^{1/3})$. These produce (1.18). Yet equation (1.18), which may seem an improvement on (1.17), still tells us very little.

Neither (1.17) nor (1.18) are incorrect. Rather, neither tells us what we want to know. A more experienced modeler and analyst might observe that we have not yet sufficiently minimized the number of dimensional variables and constants. Should we not know, for instance, that the acceleration of gravity g merely shifts the position of equilibrium without changing the frequency of oscillation ω around that equilibrium? With such knowledge we would leave g

out of the list of relevant dimensional variables and constants. Then we would have 4 dimensional variables and constants and 3 dimensions, and the analysis would produce 1 dimensionless product $\omega m^{1/2} k^{-1/2}$. Consequently,

$$\omega = C \cdot \sqrt{\frac{k}{m}} \tag{1.19}$$

where C is an undetermined number.

At this point one might worry that dimensional analysis works well only when the analyst already knows the solution. Not true! Of course, more knowledge always helps. But we need not know everything in order to learn something from dimensional analysis. Dimensional analysis has other ways to protect us from monstrosities such as $m^{4/3} g / (k \rho^{1/3})$ and, thus, other ways to move us beyond the unformed and uninformative (1.18) to the optimally informed and informative (1.19). We will learn more of these "other ways" in the chapters that follow.

Essential Ideas

According to the π theorem, every dimensionally homogeneous equation can be cast as a relation among dimensionless products. Furthermore, the number of these dimensionless products N_P is equal to the number of dimensional variables and constants N_V that describe the state or process minus the minimum number of dimensions N_D in terms of which these dimensional variables and constants can be expressed, that is, $N_P = N_V - N_D$. The Rayleigh algorithm is an analytical machine for determining the dimensionless products. A dimensional model consists of the dimensional variables and constants and their dimensional formulae.

Problems

The problems illustrate developments in the text. Answers, except those given in the problem statement, are listed in the Appendix.

1.1 **Dimensionless products**. The radiant energy per volume per differential frequency interval in a cavity surrounded by walls at temperature T is ρ where

$$\rho = \frac{8 \pi v^2}{c^3} \frac{hv}{e^{hv/k_B T} - 1}$$

This *spectral energy density* ρ has the dimension of an energy divided by a volume and a frequency. The symbols c, h, and k_B are, respectively, the speed of light, Plank's constant, and Boltzmann's constant. Assume this relation is dimensionally homogeneous and that $[h] = ML^2T^{-1}$, $[k_B] = ML^2T^{-2}\Theta^{-1}$, and $[T] = \Theta$. What are the two dimensionless products π_1 and π_2 in terms of which the spectral energy density may be expressed?

1.2 **Ice skaters**. Two skaters each of mass m and speed v approach each other along straight, parallel lines. A distance r separates their lines of approach. At closest approach they catch each other's hands and begin rotating around their common center of mass. Use the Rayleigh algorithm to determine how their frequency f of rotation depends upon these variables.

1.3 **Centripetal acceleration**. An object of mass m moves in a circle of radius r with speed v. Use the Rayleigh algorithm to determine an expression for its acceleration a.

1.4 **Walking Froude number**. Consider the assumption that the normal walking speed v of a biped or quadruped is a function of the length l of its leg (considered as a physical pendulum) and the acceleration of gravity g.

   (a) Show that the mass of the animal's leg cannot enter into a dimensionless product with v, l, and g.

   (b) Show that the one dimensionless product composed of these variables is the *walking Froude number* v^2/gl.

   (c) How does the walking speed v depend on the leg length l?

   (d) Given that many humans prefer to walk at a speed of approximately $1.4 \cdot m/s$, estimate the walking Froude number of a typical human being. (You will have to estimate the human leg length l.)

   (e) Estimate your own walking Froude number by measuring your normal walking speed and your leg length.

1.5 **Vibrating wire**. A wire of length l and mass per unit length λ pulls with a force (or tension) τ on two posts between which the wire is stretched tight and to which each end is attached. Suppose the middle of the wire is plucked. Use the Rayleigh algorithm to determine how the period Δt of the wire's oscillation depends upon l, λ, and τ.

1.6 **Falling through the center of the Earth**. Suppose you drill a tunnel through the Earth's center as illustrated in Figure 1.3. Then you drop an object of mass m from rest into the tunnel. What is the time required Δt for the object's passage from one side of the Earth to the other? This duration should depend on the gravitational constant G, the density of the Earth ρ, and the Earth's radius R. It might also depend on the mass m of the object dropped. These dimensional variables and constants, their descriptions, and their dimensional formulae are collected in Table 1.3.

Table 1.3

Δt	Duration of passage	T
G	Gravitational constant	$M^{-1}L^3T^{-2}$
ρ	Density of Earth	ML^{-3}
R	Radius of Earth	L
m	Mass of object	M

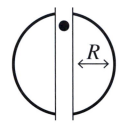

Figure 1.3. Object falling through center of Earth.

(a) Since there are 5 dimensional variables and constants expressed in terms of 3 dimensions, there should be, according to the rule of thumb $N_P = N_V - N_D$, 2 dimensionless products. Use the Rayleigh algorithm to find the two dimensionless products π_1 and π_2.

(b) There is a good reason why the mass of the object m should be eliminated from the list of variables and constants on which the duration Δt depends. Again use the Rayleigh algorithm to find the single dimensionless product implied by the variables and constants neglecting m. Derive an expression for the duration Δt.

2
Mechanics

2.1 Kinematics and Dynamics

The science of mechanics is divided into *kinematics*, the description of motion, and *dynamics*, the study of forces and their consequences. Dynamics, in turn, is divided into *equilibrium dynamics* (also known as *statics*), concerned with forces that balance each other, and *non-equilibrium dynamics*, in which forces cause bodies to accelerate. If we wish merely to describe an object in free fall, our task is kinematical. If we wish to know what determines the drag force that balances the object's weight when falling at terminal speed, our task is one of equilibrium dynamics. And finally, if we wish to know how forces cause an object to accelerate to terminal speed, our task is one of non-equilibrium dynamics. Knowing what kind of problem we face is an important part of discerning which dimensional variables and constants enter into its solution. The dimensional formulae of mechanical variables are usually expressed in terms of M, L, and T. Thus, the dimensional formula of speed is LT^{-1} and of acceleration is LT^{-2}.

Using "the dimension of" operator $[\ldots]$ on both sides of an equation is, in general, a good way to discover a dimensional formula. For instance, the dimensional formula of the dynamical quantity *force* follows, in non-equilibrium processes, from Newton's second law of motion $F = ma$. Thus, $[F] = [ma]$, $[ma] = [m][a]$, and $[m][a] = MLT^{-2}$, and, therefore, $[F] = MLT^{-2}$. Likewise, the dimensional formula of a spring constant k follows from the linear restoring force $F = -kx$ applied by a spring that is extended or compressed a distance x. Therefore, $[k] = [F]/[x]$ and so $[k] = MT^{-2}$.

2.2 Effective Dimensions

Consider a gas enclosed within a container fitted with a frictionless, massless piston of area A on which a weight of mass m rests – as shown in Figure 2.1.

Table 2.1

p	Pressure	$ML^{-1}T^{-2}$
w	Weight	MLT^{-2}
A	Area	L^2

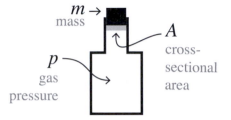

Figure 2.1. Gas in container.

Given that we ignore atmospheric pressure, how does the gas pressure p depend on the mass m? The answer, $p = mg/A$, is not in question. Rather, we are here concerned with the peculiar way dimensional analysis produces this result.

We could use 4 variables and constants, p, m, g, and A, in our dimensional analysis. But suppose we are aware that the mass m and the acceleration of gravity g enter into the result only as a single product we call the *weight* $w(= mg)$. Then we have 3 variables: p, w, and A. These symbols, their descriptions, and their dimensional formulae are collected in Table 2.1.

Since the number of dimensional variables $N_V = 3$ is the same as the number of dimensions $N_D = 3$, the Rayleigh algorithm should produce $0(= 3 - 3)$, that is, no dimensionless products – at least according to the rule of thumb $N_P = N_V - N_D$. Yet, because we know there is, indeed, one dimensionless product pA/w, we continue with the analysis.

According to the Rayleigh algorithm, the dimensionless product is found by choosing exponents α and β so that the product $pw^\alpha A^\beta$ is dimensionless, that is, so that

$$\begin{aligned} \left[pw^\alpha A^\beta\right] &= ML^{-1}T^{-2}\left(MLT^{-2}\right)^\alpha\left(L^2\right)^\beta \\ &= M^{1+\alpha}L^{-1+\alpha+2\beta}T^{-2-2\alpha} \\ &= 1. \end{aligned} \tag{2.1}$$

Therefore, we require that

$$M : 1 + \alpha = 0, \tag{2.2a}$$
$$L : -1 + \alpha + 2\beta = 0, \tag{2.2b}$$

and

$$T : -2 - 2\alpha = 0. \tag{2.2c}$$

Since equations (2.2a) and (2.2c) are algebraically equivalent, only one of them is needed. Thus, we find that $\alpha = -1$ and $\beta = 1$. As expected, this solution produces the dimensionless product pA/w or, equivalently, pA/mg and the result

$$p = C \cdot \frac{mg}{A} \tag{2.3}$$

where the dimensionless number C is undetermined by this analysis. Apparently, the Rayleigh algorithm works even when the rule of thumb $N_P = N_V - N_D$ does not.

The dimensional formulae $ML^{-1}T^{-2}$, MLT^{-2}, and L^2 listed in the rightmost column of the table reveal an interesting structure. The dimensions M and T always appear together in the combination MT^{-2}. Operationally, it is as if this coupling of M and T reduces the number of dimensions from 3 (M, L, and T) to 2 (MT^{-2} and L). Then $N_V = 3$ and $N_D = 2$, and so $N_P = N_V - N_D$ yields $N_P = 1$. This tactic saves the rule of thumb $N_P = N_V - N_D$. But why? And, does it always work?

The Number of Effective Dimensions

The rule $N_P = N_V - N_D$ is always observed if N_D is defined as *the minimum number of dimensions or combinations of dimensions needed to express the dimensional formulae of all the dimensional variables and constants.* [8] We see, for instance, that we need only 2 dimensions or combinations of dimensions, MT^{-2} and L, to express the dimensional formulae of p, w, and A. Alternatively, we need 3 dimensions or combinations of dimensions, M, L, and T, to express the dimensional formulae of the 4 dimensional variables and constants p, m, g, and A. In each case, the rule of thumb $N_P = N_V - N_D$ is observed.

For convenient reference I call "the minimum number of dimensions or combinations of dimensions needed to express the dimensional formulae of all the dimensional variables and constants" the number of *effective dimensions*. Observe that the form of the first of the two effective dimensions in this example, MT^{-2} and L, encodes some physics – and the physics encoded is not hard to identify. The mass m resting upon the piston that encloses the ideal gas is in equilibrium because the forces on it balance one another. Neither this mass nor any part of the gas is accelerating. Nowhere is Newton's second

Table 2.2

p	Pressure	$(MLT^{-2})L^{-2}$	FL^{-2}
w	Weight	(MLT^{-2})	F
A	Area	L^2	L^2

law $F = ma$ needed or invoked. Therefore, the dimensional formula of F is, in this case, disconnected from the dimensional formula of ma. What, in a more general context, has a dimensional formula of MLT^{-2} is here itself an effective dimension that we denote as the dimensional combination MLT^{-2} or more concisely as F. We depend on context to distinguish between the variable F, denoting force, and its dimension F.

Clearly only 2 dimensions, L and MLT^{-2}, that is, L and F, need appear in the dimensional formulae of p, w, and A. These symbols, their descriptions, and their dimensional formulae in terms of these effective dimensions are collected in Table 2.2. Note that 3 dimensional variables and constants and 2 effective dimensions produce, according to $N_P = N_V - N_D$, 1 dimensionless product, which we know is pA/w.

Because effective dimensions, used in this way, recognize and encode relationships among the dimensional variables and constants, their number is always less than or equal to the number of dimensions in which these variables and constants are otherwise expressed, in this case, M, L, and T. The concept of effective dimension is usually unremarked, because the Rayleigh algorithm automatically, if implicitly, identifies the number N_D of effective dimensions in such a way that $N_P = N_V - N_D$ is always observed. In particular, the number of effective dimensions is the number of independent constraints that determine the exponents defining the dimensionless products. [9] However, once identified, effective dimensions may be explicitly adopted – as we did in the last column of Table 2.2.

2.3 Imposed Dimensions

Modeling a physical state or process is more art than algorithm. One aspect of the dimensional analyst's art is to choose dimensional variables and constants that appropriately describe a physical process or state. Another is to choose the dimensions in terms of which the dimensional formulae of these variables and constants are expressed. Often, in mechanical problems we choose the dimensions M, L, and T, but in the last column of Table 2.2 we chose, instead, the dimensions L and F.

I call the dimensions in terms of which one chooses to denominate the dimensional variables and constants the *imposed dimensions* because the dimensional analyst consciously adopts or *imposes* them. [10] Imposed dimensions no less than effective dimensions define a dimensional model. As we have seen and shall see, the dimensions M, L, and T are not the only dimensions that can be imposed on a mechanical state or process.

Throughout this book, we will invariably use Rayleigh's algorithm and, therefore, will automatically take advantage of the effective dimensions implied by the list of dimensional variables and constants and the dimensions in which their formulae are expressed. On occasion, especially here in Chapter 2 and in Chapters 3 and 4, we will impose dimensions that explicitly encode relationships that further define a dimensional model even before we execute the Rayleigh algorithm. Consciously imposed dimensions may *increase* N_D and consequently *decrease* the number of dimensionless products $N_P(= N_V - N_D)$. Comparing the following analysis with the related one of Section 1.9 makes this point.

2.4 Example: Hanging Spring-Mass System

We first encountered the hanging spring-mass system in Section 1.9. As the presumably rigid mass m at the end of the spring bounces up and down, its volume V is not related to its position or to the extension or compression of the spring. Therefore, the dimensional formula of volume $[V]$ is independent of the dimension L that describes the position of the mass and the extension or compression of the spring. In particular, $[V]$ is not L^3. Rather, $[V] = V$ where, in this case, V is the symbol of an imposed dimension as well as the symbol of a dimensional variable denoting volume. Consequently, we reconfigure the table in Section 1.9 in terms of the imposed dimensions M, L, T, and V. Then the number of dimensions N_D *increases* from 3 to 4 and, consequently, the number of dimensionless products $N_P(= N_V - N_D)$ *decreases* from 3 to 2. The symbols that characterize the hanging spring-mass system, their descriptions, and their dimensional formulae in terms of these imposed dimensions are collected in Table 2.3.

The product $\omega m^\alpha k^\beta \rho^\gamma V^\delta g^\varepsilon$ becomes dimensionless when

$$
\begin{aligned}
[\omega m^\alpha k^\beta \rho^\gamma V^\delta g^\varepsilon] &= T^{-1} M^\alpha \left(MT^{-2}\right)^\beta \left(MV^{-1}\right)^\gamma V^\delta \left(LT^{-2}\right)^\varepsilon \\
&= T^{-1-2\beta-2\varepsilon} M^{\alpha+\beta+\gamma} V^{-\gamma+\delta} L^\varepsilon \\
&= 1.
\end{aligned}
\tag{2.4}
$$

Table 2.3

ω	Frequency	T^{-1}
m	Mass	M
k	Spring constant	MT^{-2}
ρ	Mass density	MV^{-1}
V	Volume	V
g	Acceleration of gravity	LT^{-2}

The constraints

$$T : -1 - 2\beta - 2\varepsilon = 0, \tag{2.5a}$$

$$M : \alpha + \beta + \gamma = 0, \tag{2.5b}$$

$$V : -\gamma + \delta = 0, \tag{2.5c}$$

and

$$L : \varepsilon = 0 \tag{2.5d}$$

are solved by $\beta = -1/2$, $\gamma = 1/2 - \alpha$, $\delta = 1/2 - \alpha$, and $\varepsilon = 0$. Therefore, $\omega m^\alpha k^\beta \rho^\gamma V^\delta g^\varepsilon = \left(\omega \rho^{1/2} V^{1/2}/k^{1/2}\right)(m/\rho V)^\alpha$ where α is arbitrary, and so the two dimensionless products are $\omega\sqrt{\rho V/k}$ and $\rho V/m$.

Because these two products are complete, we may form all others from them. In particular, if we multiply the first by the inverse square root of the second, we produce the convenient pair $\omega\sqrt{m/k}$ and $\rho V/m$. These are related by

$$\omega = \sqrt{\frac{k}{m}} \cdot h\left(\frac{\rho V}{m}\right) \tag{2.6}$$

where $h(x)$ is an undetermined function. Given $m = \rho V$, this result reduces to

$$\omega = C \cdot \sqrt{\frac{k}{m}} \tag{2.7}$$

where $C = h(1)$ is an undetermined, dimensionless number. The "monstrous" dimensionless product $m^{4/3}g/\left(k\rho^{1/3}\right)$ that plagued the analysis of Section 1.9 never emerges.

When used uncritically, as they are in Section 1.9, the dimensions M, L, and T are *absolute* because they allow, however implicitly, any and every kind of mechanical state or process. In contrast, imposed dimensions are *relative* because they restrict the particular state or process to the model adopted – with its idealizations and simplifications – rather than to the whole universe of physical possibilities.

2.5 Example: Hanging, Stretched Cable

While an idealized spring pushes or pulls at a point, materials in bulk, such as steel, glass, concrete, bone, water, or air, push or pull across an area. The equivalent "spring constant" for these materials is called the *bulk modulus*. The bulk modulus K of an isotropic material is defined as the proportionality constant between the increment in material pressure Δp and the relative increment in the material volume $\Delta V / V$ so that $\Delta p = -K(\Delta V / V)$. The negative sign indicates that a *decrease* in the volume V of the material causes an *increase* in the pressure p it exerts. Therefore, the dimensional formula of the bulk modulus $[K]$ is the same as that of pressure $[p]$. The bulk moduli of common materials in terms of their *SI* unit, the *Pascal* or *Pa,* are collected in Table 2.4.[a]

Consider a uniform, massive cable that hangs from one end. How much does the cable stretch s as a fraction of its relaxed length l? We believe that this stretch must depend not only on l but also on λ the mass per unit length of the relaxed cable, g the acceleration of gravity, and K the bulk modulus of the cable material. These symbols, their descriptions, and their dimensional formulae are collected in Table 2.5. The third column lists the dimensional formulae in terms of M, L, and T while the fourth column lists the dimensional formulae in terms of the imposed dimensions F, M, and L. This is an equilibrium problem, but introducing force F as an imposed dimension does not, in this case, increase the number of dimensions N_D. In each case, $N_V = 5$, $N_D = 3$, and, according to the rule of thumb $N_P = N_V - N_D$, the analysis should produce $2(5 - 3)$ dimensionless products.

Another tactic would be to combine mass per length λ with the acceleration of gravity g to form a weight per length λg where $[\lambda g] = FL^{-1}$. In this case the

Table 2.4

Material	Bulk moduli (in Pascals)
Steel	$160 \cdot 10^9$
Glass	$(35 - 55) \cdot 10^9$
Concrete	$30 \cdot 10^9$
Bone	$\sim 9 \cdot 10^9$
Water	$2.2 \cdot 10^9$
Air	$1.01 \cdot 10^5$

[a] These bulk moduli are at standard temperature, 0° Celsius, and standard pressure, $1.01 \cdot 10^5$ Pascals or 1 atmosphere.

Table 2.5

s	Stretch	L	L
l	Length	L	L
λ	Mass per length	ML^{-1}	ML^{-1}
g	Acceleration of gravity	LT^{-2}	FM^{-1}
K	Bulk modulus	$ML^{-1}T^{-2}$	FL^{-2}

Table 2.6

s	Stretch	L
l	Length	L
λg	Weight per length	FL^{-1}
K	Bulk modulus	FL^{-2}

table is reduced to four rows as shown in Table 2.6. This tactic reduces both N_V and N_D each by 1 and, therefore, preserves the number $N_P(=2)$ of dimensionless products. Any of these three approaches serves. Since it is easier to work with fewer variables, constants, and dimensions, we adopt Table 2.6, in which F and L are the only imposed dimensions.

In this case we do need the Rayleigh algorithm to discover the two independent and complete dimensionless products. A simple inspection of the table will do. The pair of dimensionless products s/l and $\lambda g/lK$ is particularly convenient because the variable of interest s appears in only one place. Therefore, these dimensionless products are related by

$$s = l \cdot f\left(\frac{\lambda g}{lK}\right) \tag{2.8}$$

where the function $f(x)$ is undetermined. This is as far as dimensional analysis *per se* takes us.

However opaque (2.8) may seem, it has reduced an unknown relation among 5 dimensional variables and constants to an unknown function between 2 dimensionless products. Furthermore, (2.8) is a good starting point for a different kind of analysis that, when justified, transforms an unknown function into a power of a dimensionless product.

2.6 Asymptotic Behavior

We are sometimes confronted with an undetermined function of a single variable $f(x)$ where x is either very small or very large compared to one. When

$x \ll 1$ or $x \gg 1$ there are three possibilities [11]: (1) $f(x)$ approaches a non-vanishing dimensionless constant C, (2) $f(x)$ approaches a power law $C \cdot x^n$ where n is a non-zero exponent, and (3) neither of these.

Possibility (3) includes, for instance, the function $f(x)$ oscillating indefinitely or changing exponentially or logarithmically. Possibilities (1) and (2) are summarized in the power law approximation

$$f(x) \approx C \cdot x^n \tag{2.9}$$

valid when either $x \ll 1$ or $x \gg 1$ and where both C and n are undetermined and n may vanish.

Of course, dimensional analysis alone does not reveal the asymptotic behavior of $f(x)$. But our physics sense might. If $x \ll 1$ or $x \gg 1$, we should ask ourselves, "How should $f(x)$ behave as $x \to 0$ or $x \to \infty$? Should $f(x)$ approach a finite constant or should it become monotonically small or large?" If our answer is "Yes" to either of these questions, the power law limit $f(x) \to C \cdot x^n$ may be appropriate. And, if so, we may be able to determine whether $n = 0$, $n > 0$, or $n < 0$.

To illustrate, we return to the stretched, hanging cable of Section 2.5. Suppose we are most concerned with cables of, say, a few meters, more or less, a mass density of less than one kilogram per meter, and common, solid materials with a bulk modulus on the order of $10^9 \cdot Pa$ or greater. In this case the dimensionless ratio $\lambda g / lK \leq 10^{-8}$. Since there is no reason $f(\lambda g / lK)$ should oscillate or change exponentially or logarithmically as $\lambda g / lK$ decreases, a power law asymptotic form makes sense. Therefore, we approximate (2.8) with

$$s = C \cdot l \left(\frac{\lambda g}{lK} \right)^n \tag{2.10}$$

where the number C and the exponent n are undetermined.

Our expectations for the dependence of the cable stretch s on the dimensional variables and constants, l, λ, g, and K, further constrain n. First, $n > 0$. Otherwise s would not increase with the chain's mass per unit length λ. Furthermore, because longer cables of the same composition should stretch more than shorter ones, it must be that $n < 1$. Therefore, $0 < n < 1$. A guess of $n = 1/2$ produces the scaling $s = C' \cdot \sqrt{l \lambda g / K}$. While consistent with dimensional homogeneity and our expectations, this result goes beyond what we know.

2.7 Example: Speed of Sound

Sound travels in air, water, and metals. All these are *elastic* materials whose parts, when displaced a small distance, are pushed or pulled back into position.

Table 2.7

c_S	Sound speed	LT^{-1}
K	Bulk modulus	$ML^{-1}T^{-2}$
ρ	Mass density	ML^{-3}
$m\bar{v}^2$	Internal kinetic energy	ML^2T^{-2}

The best expression of the internal forces that keep these materials close to equilibrium is their bulk modulus K.

This displacement from and return to equilibrium of one part of a material results in an oscillation whose effect is transmitted to its neighboring parts at a characteristic speed called the *sound speed*. Therefore, the sound speed c_S of an elastic material should depend on its bulk modulus K and on its mass density ρ, since the latter describes the material's inertia. What else? The internal kinetic energy of its parts $m\bar{v}^2$ as determined by a particle mass m and speed $\bar{v}$? Possibly.

The 4 dimensional variables and constants c_S, K, ρ, and $m\bar{v}^2$ describe a non-equilibrium, mechanical model. For this reason we impose the 3 dimensions M, L, and T as indicated in Table 2.7.

If the imposed dimensions M, L, and T are the effective dimensions, the Rayleigh algorithm will produce $1(4-3)$ dimensionless product. The dimensional variables and constants form a dimensionless product $c_S K^\alpha \rho^\beta (m\bar{v}^2)^\gamma$ where c_S occupies the place reserved for the variable of interest. The dimensional formula

$$\left[c_S K^\alpha \rho^\beta \left(m\bar{v}^2\right)^\gamma\right] = LT^{-1}\left(ML^{-1}T^{-2}\right)^\alpha \left(ML^{-3}\right)^\beta \left(ML^2T^{-2}\right)^\gamma$$
$$= L^{1-\alpha-3\beta+2\gamma}T^{-1-2\alpha-2\gamma}M^{\alpha+\beta+\gamma} \tag{2.11}$$

implies that when the exponents α, β, and γ satisfy

$$L: 1-\alpha-3\beta+2\gamma = 0, \tag{2.12a}$$
$$T: -1-2\alpha-2\gamma, \tag{2.12b}$$

and

$$M: \alpha+\beta+\gamma = 0 \tag{2.12c}$$

the form $c_S K^\alpha \rho^\beta (m\bar{v}^2)^\gamma$ will be dimensionless. The solution of equations (2.12) is $\alpha = -1/2$, $\beta = 1/2$, and $\gamma = 0$. Therefore,

$$c_S = C \cdot \sqrt{\frac{K}{\rho}} \tag{2.13}$$

where C is an undetermined dimensionless number. Since the bulk modulus K of an ideal gas is its pressure p, the speed of sound in an ideal gas is given by $c_S = C \cdot \sqrt{p/\rho}$.

2.8 Example: Side Window Buffeting

Many of us are familiar with the "whuppa, whuppa, whuppa," sound made when a rear window of a rapidly moving car is opened. This sound can be quite annoying – even painful. Usually the only remedy is to open more windows or to close them all. Sometimes speeding up or slowing down helps. This *side window buffeting* is more pronounced in newer, tighter, more aerodynamic cars than in older, less tight, less aerodynamic ones.

A car with a single open window is actually a large version of what is known as a *Helmholtz resonator*. When Hermann von Helmholtz (1821–1894) described their properties in 1885, he had no idea that someday human beings would climb inside large, self-propelled versions of these resonators. Helmholtz's own resonators were handheld glass or metallic cavities of volume V, usually spheroidal (but the shape is not crucial), with a small opening of area A. Helmholtz chose values of A and V so that the air inside the cavity resonated with a particular frequency ω. On the side opposite the opening of area A he added a protruding orifice with an even smaller opening that could be inserted into his ear. Helmholtz used his set of resonators to isolate and listen to particular frequencies in the sounds made by "the whistling of the wind, the rattling of carriage wheels, the splashing of water." [12] When one blows across the mouth of an empty beer bottle, one hears the low pitch to which the bottle, an accidental Helmholtz resonator, is tuned.

The resonate frequency ω of a car with an open window of area A is determined by the volume of the passenger compartment V, the mass density ρ of air, and the pressure p of the air. After all, it is the air in the car that vibrates, and the state of the air is completely determined by its volume V, its mass density ρ, and its pressure P. These symbols, their descriptions, and their dimensional formulae are collected in Table 2.8.

Table 2.8

ω	Frequency	T^{-1}
A	Area of opening	L^2
V	Resonator volume	L^3
ρ	Density of air	ML^{-3}
p	Air pressure	$ML^{-1}T^{-2}$

Because there are 5 dimensional variables, ω, A, V, ρ, and p, and 3 imposed dimensions, M, L, and T, we expect to find 2 dimensionless products related to one another by an undetermined function. These dimensionless products assume a form $\omega A^\alpha V^\beta \rho^\gamma p^\delta$ whose dimensional formula is given by

$$
\begin{aligned}
\left[\omega A^\alpha V^\beta \rho^\gamma p^\delta\right] &= T^{-1}\left(L^2\right)^\alpha \left(L^3\right)^\beta \left(ML^{-3}\right)^\gamma \left(ML^{-1}T^{-2}\right)^\delta \\
&= T^{-1-2\delta} L^{2\alpha+3\beta-3\gamma-\delta} M^{\gamma+\delta}.
\end{aligned}
\tag{2.14}
$$

The 4 exponents α, β, γ, and δ that render $\omega A^\alpha V^\beta \rho^\gamma p^\delta$ dimensionless are constrained by the three equations:

$$
T: -1 - 2\delta = 0,
\tag{2.15a}
$$

$$
L: 2\alpha + 3\beta - 3\gamma - \delta = 0,
\tag{2.15b}
$$

and

$$
M: \gamma + \delta = 0.
\tag{2.15c}
$$

Their solution, $\beta = 1/3 - 2\alpha/3$, $\gamma = 1/2$, and $\delta = -1/2$, reduces the dimensionless product to $\omega V^{1/3}(\rho/p)^{1/2}\left(A/V^{2/3}\right)^\alpha$ where α is arbitrary. Therefore,

$$
\omega = \frac{1}{V^{1/3}} \sqrt{\frac{p}{\rho}} \cdot f\left(\frac{A}{V^{2/3}}\right)
\tag{2.16}
$$

where $f(x)$ is an undetermined function.

According to (2.16), the resonant frequency ω is directly proportional to $\sqrt{p/\rho}$, which itself is proportional to the speed of sound in air. Also resonators with the same shape yet of different sizes, that is, having the same ratio $A/V^{2/3}$ but with different volumes V, are such that $\omega \propto 1/V^{1/3}$.

However, we also know that $A/V^{2/3}$ is very small in a typical Helmholtz resonator. Certainly the linear size of a car's passenger compartment $V^{1/3}$ is much larger than that of an open window $A^{1/2}$. When $A/V^{2/3} \ll 1$ (2.16) is approximated by the power law

$$
\omega = \frac{C}{V^{1/3}} \cdot \sqrt{\frac{p}{\rho}} \cdot \left(\frac{A}{V^{2/3}}\right)^n
\tag{2.17}
$$

where C and n are undetermined numbers. Helmholtz determined that $n = 1/4$ in which case (2.17) reduces to

$$
\omega = C \cdot \frac{A^{1/4}}{V^{1/2}} \sqrt{\frac{p}{\rho}}.
\tag{2.18}
$$

2.9 Example: Two-Body Orbits

The moon orbits the earth, the earth orbits the sun, and the stars in a double star system orbit each other. Each of these systems contains two bodies, one of mass m_1 and the other of mass m_2, separated by a characteristic distance r. Their gravitational attraction causes each body in a system to orbit their mutual center of mass. This motion has a period Δt after which time the massive bodies return to their initial positions and velocities. Thus, 4 dimensional variables, Δt, r, m_1 and m_2, and 1 dimensional constant, G, completely characterize this two-body, gravitationally bound system. Their symbols, descriptions, and dimensional formulae are collected in Table 2.9.

Since there are 5 dimensional variables and constants and 3 imposed dimensions, two independent and complete dimensionless products should characterize this non-equilibrium dynamical system. With practice we will develop patterns of thought that allow us to forgo the Rayleigh algorithm and quickly identify these dimensionless products. However, the Rayleigh algorithm always works and has the advantage of methodological clarity if not calculational efficiency.

Accordingly, we require that $\Delta t r^\alpha m_1^\beta m_2^\gamma G^\delta$ be dimensionless. Therefore,

$$\left[\Delta t r^\alpha m_1^\beta m_2^\gamma G^\delta\right] = TL^\alpha M^{\beta+\gamma}\left(M^{-1}L^3T^{-2}\right)^\delta$$
$$= T^{1-2\delta}L^{\alpha+3\delta}M^{\beta+\gamma-\delta}, \tag{2.19}$$

and

$$T : 1 - 2\delta = 0, \tag{2.20a}$$
$$L : \alpha + 3\delta = 0, \tag{2.20b}$$

and

$$M : \beta + \gamma - \delta = 0 \tag{2.20c}$$

renders $\Delta t r^\alpha m_1^\beta m_2^\gamma G^\delta$ dimensionless. The solution to these constraints is $\alpha = -3/2$, $\gamma = 1/2 - \beta$, and $\delta = 1/2$, and so

$$\Delta t r^\alpha m_1^\beta m_2^\gamma G^\delta = \frac{\Delta t m_2^{1/2} G^{1/2}}{r^{3/2}}\left(\frac{m_1}{m_2}\right)^\beta \tag{2.21}$$

Table 2.9

Δt	Period	T
r	Characteristic separation	L
m_1	Mass 1	M
m_2	Mass 2	M
G	Gravitational constant	$M^{-1}L^3T^{-2}$

where β is arbitrary. Therefore, the two independent dimensionless products are $\Delta t^2 m_2 G/r^3$ and m_1/m_2. These are related to one another by

$$\Delta t^2 = \frac{r^3}{m_2 G} \cdot f\left(\frac{m_1}{m_2}\right) \tag{2.22}$$

where the function $f(x)$ is undetermined. This result expresses Kepler's third law, according to which $\Delta t^2 \propto r^3$ where r is directly proportional to the semi-major axis of a planet's elliptical orbit around the sun. This is as far as dimensional analysis takes us.

Symmetry considerations take us a little further. Replacing m_1 with m_2 and m_2 with m_1 should leave (2.22) unchanged. Therefore,

$$\frac{1}{m_2} f\left(\frac{m_1}{m_2}\right) = \frac{1}{m_1} f\left(\frac{m_2}{m_1}\right) \tag{2.23}$$

or, equivalently,

$$f(x) = \frac{1}{x} f\left(\frac{1}{x}\right) \tag{2.24}$$

where x is any positive dimensionless number. Equation (2.24) is a functional equation one solution of which on the domain $x > 0$ is $f(x) = C/(1+x)$ where C is an undetermined dimensionless number. An analysis based on Newton's second law leads to $f(x) = 4\pi^2/(1+x)$ and therefore to

$$\Delta t^2 = \left[\frac{4\pi^2}{G(m_1 + m_2)}\right] r^3. \tag{2.25}$$

Essential Ideas

The dimensions in terms of which the dimensional formulae of the dimensional variables and constants are expressed are not absolute, but rather relative to the model adopted. These relative or *imposed dimensions* encode idealizations and simplifications that define the model. The number of *effective dimensions* is the minimum number of imposed dimensions or combinations of imposed dimensions in terms of which the dimensional variables and constants are expressed. The rule of thumb $N_P = N_V - N_D$ is always observed when N_D is the number of effective dimensions.

Problems

2.1 **Gravitational constant**. Use Newton's law of gravitation $F = Gm_1m_2/r^2$ to determine the dimensional formula of the gravitational constant G in terms of M, L, and T.

2.2 **Interior pressure**. A non-rotating, self-gravitating, spherical body has a mass density ρ and radius R. [In parts (a) and (b) below use the result of Problem 2.1.]

    (a) Use the Rayleigh algorithm to show that $p_o = C \cdot G\rho^2 R^2$ where p_o is the pressure at the center of the sphere and C is an undetermined dimensionless constant.

    (b) Also determine an expression for the pressure $p(r)$ at a distance r from the center of the sphere.

2.3 **Gravitational instability**. A self-gravitating, spherical body of mass density ρ and radius R rotates with at an angular frequency ω. Find how the critical frequency $\omega*$ at which the sphere begins to break up depends on ρ, R, and G. [13]

2.4 **Escape velocity**. Determine how the speed v required for any object to escape from a planet depends on the planet's mass M and radius R. See Figure 2.2.

2.5 **Transverse waveform**. A long wire with mass per unit length ρ is tightly strung between two posts as illustrated in Figure 2.3. Its tension τ is essentially uniform throughout. The wire is plucked near one end and a transverse waveform propagates as shown in Figure 2.3. Determine an expression for the waveform's speed of propagation v in terms of τ and ρ.

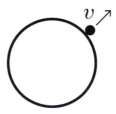

Figure 2.2. Escape velocity.

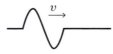

Figure 2.3. Transverse waveform.

2.6 **Drag force in an ideal fluid**. A sphere of radius r moves through a fluid of mass density ρ at a speed v. Assume the fluid is merely pushed out of the way by the sphere without, in any way, adhering to it.

    (a) How does the drag force F_D that resists this motion depend on these variables?

    (b) Suppose that the sphere is falling at terminal speed through the fluid. Given that the drag force F_D perfectly balances the sphere's weight mg, find how the terminal speed v depends on mg, ρ, and r.

2.7 **Free vibrations**. A body of mass m, size l, and bulk modulus K vibrates with a natural (unforced) frequency ω. Find an expression for ω in terms of m, l, and K.

2.8 **Soap bubble**. The air pressure within a spherical soap bubble of radius r and surface tension σ exceeds that outside the bubble by Δp. Since the soap bubble is in an equilibrium state, force is an imposed dimension F independent of M, L, and T.

    (a) How is Δp related to r and the surface tension σ of the bubble? One should find the, possibly, counterintuitive result that $\Delta p \propto r^{-1}$.

    (b) Determine N_V, N_D, and N_P. Show that $N_P = N_V - N_D$.

2.9 **Blast wave**. Energy E is quickly released at a point near the surface of the Earth. The resulting blast wave propagates through air of density ρ and creates a hemispherical shock of radius R at time t. How does R depend on time t, ρ, and E? [14]

3

Hydrodynamics

3.1 Fluid Variables

A fluid is a continuously distributed material characterized by a mass density ρ, a speed v, and, when *isotropic* or non-directional, a single pressure p at a point. In general, these basic kinematical descriptors ρ, v, and p may vary from place to place. Here we consider only incompressible fluids whose mass density ρ is constant throughout. Under normal conditions water is the most familiar realization of an incompressible fluid.

The adjacent parts of a fluid can shift with respect to one another and push, pull, and accelerate each other with forces originating in non-uniform pressure p, in viscosity μ, and in surface tension σ – as explained later in this chapter. Gravity, which tends to straighten out the free surface of a fluid by pulling its parts down toward the center of the earth, also operates.

These terms (mass density ρ, speed v, pressure p, viscosity μ, surface tension σ, and the acceleration of gravity g) are related to one another by equations of motion that, because of their many applications, have been carefully studied. Our concern is with the constraints that dimensional homogeneity imposes on the dimensional variables and constants that enter into these equations.

3.2 Example: Water Waves

Water is a relatively dense fluid whose free surface, say, on a pond or lake, is flat – at least when that surface is in equilibrium. When the surface is disturbed, gravity pulls the water back toward flatness, the water overshoots its equilibrium position, disturbs neighboring regions, and the disturbance propagates. Figure 3.1 illustrates the variables of a water wave disturbance: wavelength λ, wave amplitude a, depth of the water d, and propagation speed v.

Table 3.1

v	Wave speed	LT^{-1}
λ	Wavelength	L
a	Wave amplitude	L
d	Fluid depth	L
ρ	Mass density	ML^{-3}
g	Acceleration of gravity	LT^{-2}

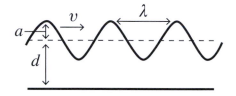

Figure 3.1. The variables of a wave disturbance on the free surface of an incompressible fluid.

Since the restoring force is gravitational, the fluid mass density ρ and the acceleration of gravity g also play a role. How does the wave speed v depend on λ, a, d, ρ, and g?

These 6 dimensional variables and constants, their descriptions, and their dimensional formulae are collected in Table 3.1. Because we express their dimensional formulae in terms of 3 imposed dimensions, M, L, and T, the Rayleigh algorithm should produce 3 dimensionless products. An inspection of Table 3.1 reveals them to be $v/\sqrt{g\lambda}$, λ/d, and a/d or, if one prefers, $v/\sqrt{g\lambda}$, $v/\sqrt{gd}$, and $v/\sqrt{ga}$. Note that the mass density ρ cannot and does not enter into these dimensionless products. Dimensional analysis takes us no further.

To go further we appeal to the approximations of wave theory and to the properties of water waves. In particular, we limit our analysis to *small amplitude waves* whose amplitude a is negligibly small compared to both the wavelength λ and the water depth d. This approximation is, of course, not valid for waves that crash on the beach, but otherwise works well. With wave amplitude a eliminated from the analysis, we have only 5 dimensional variables and constants v, λ, g, ρ, and d expressed in terms of 3 imposed dimensions M, L, and T. The 2 remaining dimensionless products, $v/\sqrt{g\lambda}$ and λ/d, are related to one another by

$$v = \sqrt{g\lambda} \cdot f\left(\frac{\lambda}{d}\right) \qquad (3.1)$$

where $f(x)$ is undetermined. However, (3.1) tells us only that the speed v of a small-amplitude water wave is proportional to the square root of the acceleration of gravity g.

The limiting cases of *deep-water waves* for which $d \gg \lambda$ and for *shallow-water waves* for which $d \ll \lambda$ yield to further analysis. In both cases (3.1) reduces to an asymptotic form. Assuming this asymptotic form is that of a power law, (3.1) becomes

$$v = C \cdot \sqrt{g\lambda} \cdot \left(\frac{\lambda}{d}\right)^n \qquad (3.2)$$

where C and n are undetermined and may be different in the two cases $\lambda/d \ll 1$ and $\lambda/d \gg 1$.

Because deep-water waves on the surface of the ocean seem to be independent of ocean depth, it must be that in this limit $n = 0$ and therefore the speed of a deep-water wave

$$v = C \cdot \sqrt{\lambda g}. \qquad (3.3)$$

Thus, longer-wavelength, deep-water ocean waves propagate more quickly than do shorter-wavelength ones. We may have observed this phenomenon firsthand or in a film of a long-wavelength swell passing under and proceeding beyond shorter-wavelength waves on the swell's surface. Further analysis indicates that, in this case, $C = 1/\sqrt{2\pi}$.

Shallow-water ocean waves for which $d \ll \lambda$ are called *tsunamis* after the Japanese word for *harbor waves*. Tsunamis, which are initiated when a large volume of ocean water is quickly displaced from equilibrium – say, by earthquake, underwater landslide, or asteroid impact – can have wavelengths as long as 1,000 kilometers. A different kind of analysis, as well as observation, indicates that tsunami speeds are independent of wavelength. In this case $n = -1/2$ and so

$$v = C' \cdot \sqrt{gd}. \qquad (3.4)$$

It has been shown that $C' = 1/\sqrt{2}$. The wave speed of the devastating tsunami of 2004 was as high as 1,000 km/hour.

3.3 Surface Tension

The force that restores the surface of a liquid to equilibrium on a small scale is *surface tension* rather than gravity. Surface tension originates in the attraction of the individual molecules of liquid to nearby molecules. Within the body of a liquid a single molecule is pulled indifferently in all directions and the net attraction vanishes. However, these attractions pull a molecule on the free surface of a liquid inward – toward the body of the fluid. It is for this reason that while large bodies of liquid, under the dominant influence of gravity, have a flat surface, small drops of liquid, under the dominant influence of surface tension, are spherical.

The surface tension of a liquid is characterized by a dimensional quantity σ defined as the net force f that the molecules on one side of a short line on the surface of the liquid exert on the adjacent molecules on the opposite side of that line, divided by the length l of the line. Thus, $\sigma = f/l$ and $[\sigma] = FL^{-1}$ or $[\sigma] = MT^{-2}$ depending on whether these forces are balanced or not.

Capillary Waves

Capillary waves are small surface waves or ripples whose dominant restoring force is surface tension. Suppose a liquid is indefinitely deep and the amplitude of a capillary wave is so small as to play no role in its behavior. Then its wave speed v can depend only on wavelength λ, surface tension σ, acceleration of gravity g, and, possibly, the mass density ρ of the liquid. These 5 symbols, their descriptions, and their dimensional formulae are collected in Table 3.2.

Because the dimensional formulae of these 5 dimensional variables and constants are expressed in terms of 3 imposed dimensions, M, L, and T, the variables and constants should combine into 2 independent, dimensionless products. These take a form whose dimensional formula,

$$
\begin{aligned}
\left[v\lambda^{\alpha}\sigma^{\beta}g^{\gamma}\rho^{\delta}\right] &= \left(LT^{-1}\right)L^{\alpha}\left(MT^{-2}\right)^{\beta}\left(LT^{-2}\right)^{\gamma}\left(ML^{-3}\right)^{\delta}, \\
&= L^{1+\alpha+\gamma-3\delta}T^{-1-2\beta-2\gamma}M^{\beta+\delta}
\end{aligned}
\tag{3.5}
$$

implies that the exponents α, β, γ, and δ are constrained by

$$
L : 1 + \alpha + \gamma - 3\delta = 0,
\tag{3.6a}
$$

$$
T : -1 - 2\beta - 2\gamma = 0,
\tag{3.6b}
$$

and

$$
M : \beta + \delta = 0.
\tag{3.6c}
$$

Table 3.2

v	Wave speed	LT^{-1}
λ	Wavelength	L
σ	Surface tension	MT^{-2}
g	Acceleration of gravity	LT^{-2}
ρ	Mass density	ML^{-3}

The solution to equations (3.6) is $\beta = -1/4 - \alpha/2$, $\gamma = -1/4 + \alpha/2$, and $\delta = 1/4 + \alpha/2$. Consequently, the dimensionless products are $v[\rho/(\sigma g)]^{1/4}$ and $\lambda(g\rho/\sigma)^{1/2}$. By multiplying the first by the square root of the second, we produce $v(\lambda\rho/\sigma)^{1/2}$. This dimensionless product and $\lambda(g\rho/\sigma)^{1/2}$ compose an independent pair. In this way we find that the wave speed

$$ v = \sqrt{\frac{\sigma}{\lambda\rho}} \cdot f\left(\lambda\sqrt{\frac{g\rho}{\sigma}}\right) \tag{3.7} $$

where $f(x)$ is undetermined by our analysis.

The size of the dimensionless argument $\lambda\sqrt{g\rho/\sigma}$ of the function $f(x)$ determines whether it is gravity or surface tension that dominates. When gravity dominates $\lambda\sqrt{g\rho/\sigma} \gg 1$, it must be that $f\left(\lambda\sqrt{g\rho/\sigma}\right) \to C \cdot \lambda\sqrt{g\rho/\sigma}$ in order to recover (3.3), $v = C \cdot \sqrt{\lambda g}$, the latter derived under the assumption that $\sigma = 0$. When surface tension dominates $\lambda\sqrt{g\rho/\sigma} \ll 1$, it must be that $f\left(\lambda\sqrt{g\rho/\sigma}\right) \to f(0)$ in order that the acceleration of gravity g vanish from the result. In this case, $v = C' \cdot \sqrt{\sigma/\lambda\rho}$ where $C' = f(0)$. When $\lambda\sqrt{g\rho/\sigma} \approx 1$, both gravity and surface tension play a role.

The distance $\sqrt{\sigma/g\rho}$ that separates these two regimes is easily calculated for water. Water at 25 °C and atmospheric pressure has a surface tension $\sigma = 0.0720 \cdot N/m$ and density $\rho = 0.997 \cdot 10^3 \cdot kg/m^3$. Given $g = 9.8 \cdot m/s^2$, we find that $\sqrt{\sigma/g\rho} = 2.7 \cdot 10^{-3} \cdot m$ or $2.7 \cdot mm$. In general, pure capillary waves have wavelengths so small ($\lambda \ll 2.7 \cdot mm$) as to be invisible to the naked eye. Visible waves with wavelengths of a few millimeters up to a centimeter are mixed capillary and gravity waves.

3.4 Example: Largest Water Drop

The distance $\sqrt{\sigma/g\rho}(= 2.7 \cdot mm)$ also determines the size of the largest drop of water. Imagine a drop that hangs from a water faucet and is very slowly filled. Eventually the drop becomes so heavy it detaches from the faucet and falls. What is the volume V of this largest drop? Certainly V depends on σ, for without surface tension no drop would form. The mass density of water ρ and the acceleration of gravity g must also play a role, because it is only because ρ and g are non-vanishing that the drop detaches and falls. These dimensional variables and constants, their descriptions, and their dimensional formulae are collected in Table 3.3.

Since this is an equilibrium problem, we are free to impose force F as a dimension. In this case, the variables, descriptions, and formulae are given in

Table 3.3

V	Volume	L^3
σ	Surface tension	MT^{-2}
ρ	Mass density	ML^{-3}
g	Acceleration of gravity	LT^{-2}

Table 3.4

V	Volume	L^3
σ	Surface tension	FL^{-1}
w	Weight density	FL^{-3}

Table 3.4. Note that the dimension F enters into those dimensional formulae that express counterbalancing forces.

Both sets of variables and formulae produce the same dimensionless product. Since it is slightly easier to work with 3 rather than with 4 dimensional variables and constants, we do so here. We need no special algorithm to discover, from the second table, the single dimensionless product produced: $V(w/\sigma)^{3/2}$. Therefore, given $w = \rho g$, we find that

$$V = C \cdot \left(\frac{\sigma}{\rho g}\right)^{3/2} \tag{3.8}$$

where C is an undetermined dimensionless number. Here $(\sigma/g\rho)^{3/2} = 2.0 \cdot 10^{-8} m^3$ or $20 \cdot mm^3$.

Dimensional analysis does not determine the number C in (3.8). But C can be empirically determined. Catching and counting the water drops falling slowly from my kitchen faucet and measuring their total volume yields $C \approx 9$ so that

$$V \approx 9 \cdot \left(\frac{\sigma}{\rho g}\right)^{3/2}$$
$$\approx 180 \cdot mm^3. \tag{3.9}$$

Therefore, each of these drops has a mass of about 180 milligrams, a volume of about $180 \cdot mm^3$, and a radius of about $5.6 \cdot mm$. These are the largest drops that can be formed under the circumstance described. Smaller drops can, of course, be formed, say, by a nozzle that mixes air and water, and larger drops could be formed in a weightless environment.

3.5 Viscosity

Among the variables and constants that describe an incompressible fluid, the *viscosity* μ is likely to be the least familiar. The viscosity of a fluid quantifies its tendency, through internal friction, to reduce the relative motion of its adjacent parts. Without viscosity, eddies in water would circulate forever. All fluids are, to a degree, viscous, but viscosity is not always important in all phenomena.

Imagine a fluid trapped between two large, flat, horizontal plates. One plate is fixed in place and the other moves horizontally at constant speed. The fluid resists the motion of the moving plate with a drag force F_D per unit area A that is proportional to its speed v and inversely proportional to the separation y of the two plates. Symbolically,

$$\frac{F_D}{A} \propto \frac{v}{y}. \tag{3.10}$$

The *dynamic* or *shear viscosity* μ is the proportionality constant that turns (3.10) into the equality

$$\frac{F_D}{A} = \mu \frac{v}{y}. \tag{3.11}$$

Therefore, $[\mu] = ML^{-1}T^{-1}$ or, in equilibrium problems, $[\mu] = FL^{-2}T$. In practice, measurements of viscosity are often carried out on a thin layer of fluid trapped between two closely spaced, concentric cylinders, one of which can rotate while the other is attached to a torsion meter.

Terminal Speed

Since viscosity quantifies the ability of a fluid to resist the motion of an object passing through it, viscosity contributes to the phenomenon of *terminal speed*. Objects falling in a fluid, such as air, will approach a terminal speed that depends on the weight and shape of the object and the properties of the fluid – particularly its viscosity.

Because objects fall relatively quickly in air, the Greek philosopher Aristotle (384–322 BCE) may have developed his ideas about falling objects by extrapolating from a study of the descent of stony pebbles in a clear pool of water. [15] According to Aristotle, all space is *full*, that is, filled with a fluid, whether water or air, or something less dense but akin to these fluids. Thus, Aristotle imagined that descent in water is structurally similar to descent in air.

The dimensional analysis of a spherical pebble falling at terminal speed through a viscous fluid employs the symbols, descriptions, and dimensional formulae found in Table 3.5. Since this is an equilibrium problem, the fourth

Table 3.5

v	Terminal speed	LT^{-1}	LT^{-1}
μ	Viscosity	$ML^{-1}T^{-1}$	$FL^{-2}T$
mg	Weight	MLT^{-2}	F
r	Radius	L	L

column includes the force as an imposed dimension F. In either case there are 4 dimensional variables and constants and 3 imposed dimensions. Therefore, the analysis should produce 1 dimensionless product.

Adopting the imposed dimensions M, L, and T, we find that

$$\left[v\mu^{\alpha}(mg)^{\beta}r^{\gamma} \right] = LT^{-1}\left(ML^{-1}T^{-1}\right)^{\alpha}\left(MLT^{-2}\right)^{\beta}L^{\gamma} \qquad (3.12)$$
$$= L^{1-\alpha+\beta+\gamma}T^{-1-\alpha-2\beta}M^{\alpha+\beta}$$

where the exponents must satisfy

$$L : 1 - \alpha + \beta + \gamma = 0, \qquad (3.13a)$$
$$T : -1 - \alpha - 2\beta = 0, \qquad (3.13b)$$

and

$$M : \alpha + \beta = 0 \qquad (3.13c)$$

in order to make $v\mu^{\alpha}(mg)^{\beta}r^{\gamma}$ dimensionless. The solution to these constraints is $\alpha = 1$, $\beta = -1$, and $\gamma = 1$. Therefore, the single dimensionless product is $v\mu r/mg$, and so

$$v = C \cdot \frac{mg}{\mu r} \qquad (3.14)$$

where C is an undetermined, dimensionless number. The terminal speed of an object falling in a viscous fluid is, indeed, directly proportional to the object's weight mg and inversely proportional to the fluid's resistance, that is, to its viscosity μ – just as Aristotle claimed for all falling objects. [16]

Stokes' Law

At terminal speed the drag force F_D the fluid exerts on a spherical object by virtue of the fluid's viscosity balances the object's weight mg. Given (3.14), this drag force

$$F_D = C' \cdot v r \mu \qquad (3.15)$$

where $C' = 1/C$. Relation (3.15) with $C' = 6\pi$ is known as *Stokes' law* after George Stokes (1819–1903), the British mathematician, scientist, and engineer who discovered it in 1851.

3.6 Example: Hydraulic Jump

Anyone with a kitchen sink has observed the *hydraulic jump*. As water falls from the faucet and strikes the relatively flat bottom of the sink, it spreads out in a smooth circular pattern that is surrounded by a stationary wave form or *jump*. Beyond the jump, the water flows more deeply and roughly. This jump develops in order to maintain a constant rate of flow given that viscosity continually slows the water flowing along the sink bottom. Figure 3.2 illustrates the situation.

The radial position R at which the hydraulic jump forms depends on the mass density ρ of water and its viscosity μ. Because the water's volumetric flow rate $\pi r^2 v$ is also important, the radius r of the bottom of the cylindrical column of falling water and the speed v with which the water at the bottom of the column strikes the sink are also determinants. (Because $\pi r^2 v$ is constant along the column, it narrows as the falling water picks up speed.) These symbols, their descriptions, and their dimensional formulae are collected in Table 3.6.

Since there are 5 dimensional variables and constants and 3 imposed dimensions, we expect, according to the rule of thumb $N_P = N_V - N_D$, to find 2 dimensionless products. These have the form $R\rho^\alpha \mu^\beta v^\gamma r^\delta$. Thus, their dimensional formulae are given by

$$
\begin{aligned}
\left[R\rho^\alpha \mu^\beta v^\gamma r^\delta\right] &= L\left(ML^{-3}\right)^\alpha \left(ML^{-1}T^{-1}\right)^\beta \left(LT^{-1}\right)^\gamma L^\delta \\
&= L^{1-3\alpha-\beta+\gamma+\delta} M^{\alpha+\beta} T^{-\beta-\gamma}
\end{aligned}
\tag{3.16}
$$

Table 3.6

R	Position of jump	L
ρ	Mass density	ML^{-3}
μ	Viscosity	$ML^{-1}T^{-1}$
v	Speed	LT^{-1}
r	Radius of column	L

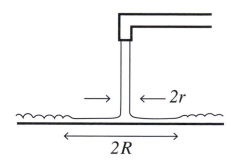

Figure 3.2. Hydraulic jump.

where the exponents α, β, γ, and δ are constrained by

$$L : 1 - 3\alpha - \beta + \gamma + \delta = 0, \tag{3.17a}$$

$$M : \alpha + \beta = 0, \tag{3.17b}$$

and

$$T : -\beta - \gamma = 0. \tag{3.17c}$$

The solution of (3.17) is $\beta = -\alpha$, $\gamma = \alpha$, and $\delta = -1 + \alpha$. These determine two dimensionless products, R/r and $r\rho v/\mu$, that are related by

$$\frac{R}{r} = f\left(\frac{r\rho v}{\mu}\right) \tag{3.18}$$

where $f(x)$ is undetermined.

Reynolds Number

The dimensionless ratios $r\rho v/\mu$ and $R\rho v/\mu$ are examples of *Reynolds numbers*, so called after Osborne Reynolds (1842–1912) who promoted their use. In general, the size of a Reynolds number in a given geometry determines whether a fluid in that geometry is dominated by viscosity or not. When the Reynolds number is small, viscosity dominates and the flow is *smooth*, *streamline*, or *laminar*. When the Reynolds number is large, viscosity is not so important and the flow is *rough* or *turbulent*. The critical Reynolds number separating these two regimes is not necessarily 1, but rather depends on the geometry of the flow. When water flows over a flat surface, as it does along the sink bottom, the relevant, critical Reynolds number $R\rho v/\mu$ is, in fact, about $5 \cdot 10^5$.

Numbers that characterize the flow of water in my sink are $R = 0.20 \cdot m$, $r = 0.02 \cdot m$, and $v = 2 \cdot m/s$. Given that for water $\rho = 10^3 \cdot kg/m^3$ and $\mu = 8.9 \cdot 10^{-4} \cdot kg/(m \cdot s)$, I find that $R\rho v/\mu = 4.5 \cdot 10^5$ – a value close to the critical Reynolds number for this geometry. Because $r\rho v/\mu$ is also large compared to 1, we replace $f(r\rho v/\mu)$ on the right-hand side of (3.18) with its power law approximation. Thus,

$$\frac{R}{r} = C \cdot \left(\frac{r\rho v}{\mu}\right)^n \tag{3.19}$$

where C and n are undetermined dimensionless numbers.

Experience helps us determine the possible values of n. First, we rewrite (3.19) as

$$R = C \cdot r^{1-n}\left(\frac{\rho r^2 v}{\mu}\right)^n. \tag{3.20}$$

I observe that a larger flow rate $\rho r^2 v$ causes a larger jump position R. Therefore, $n > 0$. I also observe that increasing r while preserving $\rho r^2 v$, say, by raising a cookie sheet under a steady stream of water, weakly increases R. Therefore, $n < 1$ and so $0 < n < 1$. According to one model for which there is experimental support, $n = 1/3$. [17] In this case (3.20) becomes

$$R = C \cdot r \left(\frac{\rho r v}{\mu} \right)^{1/3}. \tag{3.21}$$

Given the above data, I find that, for my sink and faucet, $C \approx 0.25$.

3.7 Example: Equilibrium Pipe Flow

Consider the equilibrium flow of an incompressible fluid at a volumetric rate Q from left to right through a straight, horizontal pipe of length l and uniform cross-sectional area A as illustrated in Figure 3.3. The quantity Q/A is the average fluid speed v over a cross-section. Thus, $Q = vA$. By *equilibrium flow* I mean one for which the net force on each part of the fluid, that is, on each *fluid element*, vanishes. Therefore, the velocity of each fluid element is constant.

Given the interaction with the walls of the pipe and the retarding effect of the fluid's viscosity μ, a force must push the fluid from left to right in order to keep it moving at constant speed. This force is maintained by a pressure gradient ∇p pointing in the direction opposite to the flow, that is, from right to left. Therefore, as the fluid moves from left to right over length l, its pressure drops by an increment $l\nabla p$. How does the flow rate Q depend on the cross-sectional area A, the pressure gradient ∇p, the fluid viscosity μ, and the fluid mass density ρ?

Since this is an equilibrium problem, the dimension F is an appropriate imposed dimension along with and independent of M, L, and T. As usual, the

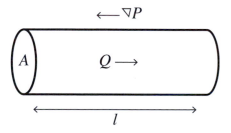

Figure 3.3. Pipe flow variables.

Table 3.7

Q	Flow rate	$L^3 T^{-1}$
∇p	Pressure gradient	FL^{-3}
μ	Viscosity	$FL^{-2}T$
ρ	Mass density	ML^{-3}
A	Cross-sectional area	L^2

dimension F enters only into those dimensional formulae that represent counterbalancing forces. In particular, $[\Delta p] = FL^{-3}$ and $[\mu] = FL^{-2}T$.

Note that imposing the dimension F, as we do here, is mathematically distinct from using the relationship $[F] = [m][a]$ and its consequence $F = MLT^{-2}$ to eliminate one of the dimensions M, L, or T in favor of F. The former restricts the solution to one in which the net force on every fluid element vanishes. In this case, forces are independent of mass times acceleration. The latter applies Newton's second law $F = ma$ by adopting the equivalence $F = MLT^{-2}$.

The dimensional variables and constants, their descriptions, and their dimensional formulae are collected in Table 3.7. Since there are 5 dimensional variables and constants ($Q, \nabla p, \mu, \rho$, and A) and 4 imposed dimensions (M, L, T, and F), there should be 1 dimensionless product. Table 3.7 reveals that, since the dimension M appears only once, the mass density ρ cannot enter into this dimensionless product.

Therefore, the dimensionless product assumes a form $Q(\nabla P)^{\alpha}\mu^{\beta}A^{\gamma}$ where the exponents, α, β, and γ, make

$$[Q(\nabla P)^{\alpha}\mu^{\beta}A^{\gamma}] = L^3 T^{-1}\left(FL^{-3}\right)^{\alpha}\left(FL^{-2}T\right)^{\beta}\left(L^2\right)^{\gamma} \qquad (3.22)$$
$$= L^{3-3\alpha-2\beta+2\gamma}T^{-1+\beta}F^{\alpha+\beta}$$

equal to 1. Thus,

$$L : 3 - 3\alpha - 2\beta + 2\gamma = 0, \qquad (3.23a)$$

$$T : -1 + \beta = 0, \qquad (3.23b)$$

and

$$F : \alpha + \beta = 0. \qquad (3.23c)$$

The solution to equations (3.23) is $\alpha = -1$, $\beta = 1$, and $\gamma = -2$. Therefore, the single dimensionless product is $Q\mu/\left(A^2 \nabla p\right)$ and so

$$Q = C \cdot \frac{A^2 \nabla p}{\mu} \qquad (3.24)$$

where C is undetermined. As seems reasonable, the volumetric flow rate Q increases with increased cross-sectional area A, increases with increased pressure gradient ∇p, and decreases with increased viscosity μ.

3.8 Example: Non-Equilibrium Pipe Flow

If we allow parts of the fluid to accelerate, then F is no longer an appropriate imposed dimension and should, in the above table in Section 3.7, be replaced with MLT^{-2}. The result is shown in Table 3.8. Since there are now only 3 imposed dimensions in terms of which the original 5 dimensional variables and constants are expressed, the Rayleigh algorithm should produce 2 dimensionless products. One of these, $Q\mu/(A^2\nabla p)$, has already been determined.

The dimensionless products assume a form $Q(\nabla p)^\alpha \mu^\beta \rho^\gamma A^\delta$ whose dimensional formula

$$[Q(\nabla p)^\alpha \mu^\beta \rho^\gamma A^\delta] = L^3 T^{-1}(ML^{-2}T^{-2})^\alpha (ML^{-1}T^{-1})^\beta (ML^{-3})^\gamma (L^2)^\delta \tag{3.25}$$
$$= L^{3-2\alpha-\beta-3\gamma+2\delta}T^{-1-2\alpha-\beta}M^{\alpha+\beta+\gamma}$$

equals 1. Therefore,

$$3 - 2\alpha - \beta - 3\gamma + 2\delta = 0, \tag{3.26a}$$
$$-1 - 2\alpha - \beta = 0, \tag{3.26b}$$

and

$$\alpha + \beta + \gamma = 0. \tag{3.26c}$$

The solution to (3.26), $\beta = -1 - 2\alpha$, $\gamma = 1 + \alpha$, and $\delta = -1/2 + 3\alpha/2$, produces two dimensionless products: $Q\rho/(\mu A^{1/2})$ and $\rho A^{3/2}\nabla p/\mu^2$. Multiplying the first by the inverse of the second yields the dimensionless product $Q\mu/(A^2\nabla p)$ that, according to Section 3.7, characterizes equilibrium flow in a straight pipe. The relation between this dimensionless product and $Q\rho/(\mu A^{1/2})$,

Table 3.8

Q	Flow rate	$L^3 T^{-1}$
∇p	Pressure gradient	$ML^{-2}T^{-2}$
μ	Viscosity	$ML^{-1}T^{-1}$
ρ	Mass density	ML^{-3}
A	Cross-sectional area	L^2

$$Q = \frac{A^2 \nabla p}{\mu} \cdot f\left(\frac{Q\rho}{\mu A^{1/2}}\right), \tag{3.27}$$

where the function $f(x)$ is undetermined, characterizes non-equilibrium flow in a straight pipe.

Since $Q = Av$, the argument $Q\rho/(\mu A^{1/2})$ is the Reynolds number $v\rho A^{1/2}/\mu$ of this flow. We recover equilibrium flow (3.24) from non-equilibrium flow (3.27) by making the fluid mass density ρ negligibly small compared with $\mu A^{1/2}/Q$ or, equivalently, by making the viscosity μ very large compared to $Q\rho/A^{1/2}$. These moves are equivalent to replacing the function $f\left(Q\rho/\mu A^{1/2}\right)$ with its low Reynolds number asymptotic value $f(0)$ which produces (3.24) with $C = f(0)$. Apparently, a vanishing Reynolds number, in a straight pipe, means the fluid moves down the pipe at constant velocity.

The general result (3.27) also has a high Reynolds number, asymptotic limit, describing *fully developed turbulent flow*, in which viscosity μ becomes negligible compared to $Q\rho/A^{1/2}$. In order that viscosity be completely eliminated from (3.27) it must be that $f(x) \to C'/x$ when $x \gg 1$. Therefore, in this fully developed turbulent flow limit $Q \to C' \cdot A^{5/2} \nabla p/(Q\rho)$ or, equivalently,

$$Q = C'' \cdot \left(\frac{\nabla p}{\rho}\right)^{1/2} A^{5/4} \tag{3.28}$$

where $C'' = \sqrt{C'}$.

3.9 Scale Models

The result of a dimensional analysis is often that one dimensionless product equals an unknown function of another. Relation (3.27) among the variables describing pipe flow for arbitrary Reynolds number,

$$\frac{Q\mu}{A^2 \nabla p} = f\left(\frac{Q\rho}{\mu A^{1/2}}\right), \tag{3.29}$$

is one example.

Suppose our job is to design a large pipe of cross-sectional area A and length l that transports motor oil. We want to test our design on a smaller-scale version of that pipe. But pumping motor oil through a smaller pipe with the same aspect ratio $A^{1/2}/l$ is not adequate. While proportionally decreasing $A^{1/2}$ and l preserves the geometry of the pipe, doing so does not preserve the Reynolds number $Q\rho/(\mu A^{1/2})$ of the flow, and it is the Reynolds number that determines whether the flow is laminar or turbulent. In fact, preserving *geometric similitude* is not necessary at all. Rather, what needs to be preserved, in this case, is *dynamic similitude*.

One preserves dynamic similitude by making the values of the two dimensionless products $Q\mu/(A^2\nabla p)$ and $Q\rho/(\mu A^{1/2})$ realized in the scale model identical to those in the fully designed pipe. We do this by varying the pressure gradient ∇p in the scale model and measuring the resulting flow rate Q – the other quantities being constant – and so determining the relationship between $Q\mu/(A^2\nabla p)$ and $Q\rho/(\mu\sqrt{A})$ in the scale model, that is, by determining the function $f(x)$. The pressure gradient ∇p in the scale model is then chosen so that the two dimensionless products are the same in the scale model and the fully designed pipe. Cars, ships, and planes, as well as pipes, have been designed in this way.

However, scale models should not depart from the physics incorporated in the full design. If, for instance, the scale model of a pipe is too small, surface tension becomes significant and the scaling implied by (3.29) breaks down.

Essential Ideas

Dimensional models of incompressible fluids introduce the dimensional variables mass density ρ, flow speed v, and pressure p as well as surface tension σ and viscosity μ. The Reynolds number is a dimensionless product that when relatively small produces laminar flow and when relatively high produces turbulent flow.

Problems

3.1 **Capillary waves**. The context is Sections 3.2 and 3.3. Use the Rayleigh algorithm to show that the speed v of a capillary wave with wavelength λ in an compressible fluid with mass density ρ and surface tension σ is given by $v = C \cdot \sqrt{\sigma/\lambda\rho}$ where C is an undetermined dimensionless number.

3.2 **Water drop oscillation**. Use either the Rayleigh algorithm or a less formal dimensional argument to determine how the oscillation frequency ω of a small drop of incompressible fluid in free fall depends on the volume V of the drop, its mass density ρ, and its surface tension σ.

3.3 **Capillary effect**. When a small-diameter tube open at both ends is inserted into water, as shown in Figure 3.4, the water is drawn a certain distance h up the tube. This distance depends on the diameter d of the tube, the surface tension of water σ, the mass density ρ of the water,

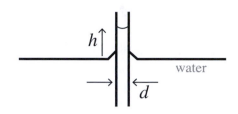

Figure 3.4. Capillary effect.

and the acceleration of gravity g. Find the 2 dimensionless products implied by these dimensional quantities and cast them into an expression for h.

3.4 **Terminal speed**. The context is the discussion of viscosity dominated terminal speed in Section 3.6. Find a dimensionally correct expression for the time t required for an object with weight mg and size r falling in a fluid of viscosity μ to reach terminal speed.

3.5 **Viscous rebound**. When a spoon is depressed into and removed from the surface of a pool of molasses, an interval Δt is required for the molasses to return to its original, equilibrium shape. (a) Find the two dimensionless products π_1 and π_2. (b) Express Δt in terms of the other variables and constants.

3.6 **Lift**. Determine a dimensionally correct expression for the mass m of a rock with mass density ρ that can be lifted from the bottom of a streambed and moved downstream. The stream is flowing with a speed v.

3.7 **Dimensionless products**. According to Section 3.1, mass density ρ, speed v, pressure p, viscosity μ, surface tension σ, and the acceleration of gravity g are the basic dimensional descriptors of hydrodynamics. Given these 6 dimensional variables and constants expressed in terms of the 3 dimensions M, L, and T, find 3 independent, dimensionless products, π_1, π_2, and π_2.

4

Temperature and Heat

4.1 Heat Transfer

Joseph Black's early work (*circa* 1760) on heat capacity and latent heat effectively disentangled the concepts of *temperature* and *heat*. Black (1728–1799) noticed, for instance, that boiling water absorbs heat without changing its temperature. The role of temperature is, apparently, to determine the direction and rate of heat transfer when two bodies, with unequal temperatures, are brought into thermal contact. Furthermore, heat moves from place to place while keeping its quantity conserved – or so it was thought.

This understanding of heat as an inherently conserved quantity remained possible until the 1840s. Then James Joule (1818–1889), in a series of increasingly precise experiments, began generating, through the dissipation of work, predictable amounts of this supposedly conserved heat. In this way Joule forged a new understanding of heat as *heating*, that is, as simply one way of transferring *energy* to or from a system. The other category of energy transfer is work done on or by a system. In 1850, Rudolph Clausius (1822–1888) harmonized this understanding of heating and working and Sadi Carnot's earlier (1826) articulation of a universal law of heat engine performance into the first and second laws of the new science of *thermodynamics*.

Thermodynamics is a theory that, while providing a new foundation for known facts, preserves the validity of older explanations in particular limits. One of these limits is that in which no work is done on or by the system in question and heat moves from place to place *as if it were a conserved quantity*. In this limit it is permissible and often convenient to use the word *heat* as short hand for "that energy transferred when no work is done." This limit is easily recognized and rich in phenomena.

The limit in which heat retains its status as an independent, conserved quantity is the regime of *heat transfer*. Calorimetry experiments are done and heat capacities are sometimes measured in this regime. [The word *calorimetry* refers to the traditional unit of heat – the *calorie*.] Interestingly, Joseph Fourier (1768–1830) first worked out the mathematics of heat transfer in the same treatise, *The Analytical Theory of Heat* (1822), in which he articulated the principle of dimensional homogeneity.

4.2 The Dimensions of Temperature and Heat

The physics of heat transfer requires two dimensions, temperature Θ and heat H, beyond those that characterize non-equilibrium dynamics. Thus, for instance,

$$\left[c_p\right] = H\Theta^{-1}M^{-1} \tag{4.1}$$

where c_p is a heat capacity per unit mass at constant pressure. We depend on context to distinguish between the traditional symbols for temperature T and the dimension time T. Thus, $[T] = \Theta$ while, if v is a velocity, $[v] = LT^{-1}$.

When work is dissipated into or produced out of the internal energy of a thermodynamic system, the first law of thermodynamics is necessarily invoked. Then the dimension heat H loses its independence and should be replaced with the dimension of energy ML^2T^{-2}. Yet when no work is done on or by a system, heat H is an appropriate imposed dimension just as when no part of a mechanical system is accelerated force F is an appropriate imposed dimension.

Occasionally, one reads that the variable temperature does not need its own dimension Θ because temperature can always be measured in energy units with dimensional formula ML^2T^{-2}. The three examples in this chapter, in Sections 4.4, 4.7, and 4.8, are counterexamples to this claim. Those in Sections 5.7, 6.2, 6.3, and 6.6 illustrate states and processes in which temperature can be measured in energy units.

4.3 Conduction and Convection

Heat conduction is accomplished when the energy of some atoms and molecules is imparted to other nearby atoms and molecules. Solids, liquids, gases, and plasmas all conduct heat. *Thermal conductivity* is a dimensional constant that characterizes a particular material's ability to conduct heat. In general, thermal conductivity may vary from point to point and even change in time as the material properties change.

The extent to which a material conducts heat at a point is proportional to the temperature gradient at that point. A one-dimensional quantification of this statement is

$$q \propto -\frac{\partial T}{\partial x} \qquad (4.2)$$

where q is the *heat flux* or rate at which heat flows past a point per unit area, T is the temperature, and $\partial T/\partial x$ is the x component of the temperature gradient. The negative sign in (4.2) ensures that the heat flux is always in a direction opposed to the temperature gradient, that is, always from hot to cold. The *thermal conductivity* k is the proportionality constant that turns (4.2) into an equality. Therefore,

$$q = -k\frac{\partial T}{\partial x} \qquad (4.3)$$

governs heat conduction.

Convection occurs when a fluid, by virtue of its bulk motion, carries its internal energy from one place to another. Convection may occur either by design, in heating and cooling systems, or naturally, as in weather patterns.

4.4 Example: Cooking a Turkey

Each year on their Thanksgiving holiday Americans consume some 50 million turkeys. Unlike chickens, which have been bred to uniform size, turkey weights vary over an order of magnitude. Wild turkey hens weigh as little as 3.0 kilograms (6.6 pounds) while domestic toms weigh as much as 36 kilograms (79 pounds). Turkeys found in grocery stores are usually between 4.5 and 11 kilograms (10–24 pounds). So the question arises, "How long should we cook our turkey? Last year our 9-kilogram turkey was perfect after 4 hours. This year we have a 6-kilogram turkey."

What are the physical variables and constants that contribute to a turkey's cooking time Δt? Certainly its mass m and possibly the difference $\Delta T (= T_{oven} - T_o)$ between the turkey's initial temperature T_o and the temperature $T_{oven} (> T_o)$ at which the oven is set are relevant. Since the turkey's thermal conductivity k determines how a temperature difference leads to a heat flux, k should also contribute to the cooking time. So should the turkey's heat capacity per unit mass at constant pressure c_p and its mass density ρ. These parameters seem sufficient. Their symbols, descriptions, and dimensional formulae are collected in Table 4.1. Note that the imposed dimension heat H appears in the dimensional formulae of the dimensional quantities that quantify heat conduction.

Table 4.1

Δt	Cooking time	T
m	Turkey mass	M
ΔT	Temperature difference	Θ
k	Thermal conductivity	$HT^{-1}L^{-1}\Theta^{-1}$
c_p	Specific heat at constant pressure	$HM^{-1}\Theta^{-1}$
ρ	Mass density	ML^{-3}

Since there are 6 dimensional variables and constants, Δt, m, ΔT, k, c_p, and ρ and 5 imposed dimensions, M, L, T, H, and Θ, our analysis should produce 1 dimensionless product. This product will assume a form $\Delta t m^{\alpha} \Delta T^{\beta} k^{\gamma} c_p^{\delta} \rho^{\varepsilon}$ with exponents, α, β, γ, δ, and ε, that render the form dimensionless. Therefore,

$$
\begin{aligned}
\left[\Delta t m^{\alpha} \Delta T^{\beta} k^{\gamma} c_p^{\delta} \rho^{\varepsilon}\right] &= TM^{\alpha}\Theta^{\beta}\left(HT^{-1}L^{-1}\Theta^{-1}\right)^{\gamma}\left(HM^{-1}\Theta^{-1}\right)^{\delta}\left(ML^{-3}\right)^{\varepsilon} \\
&= T^{1-\gamma}M^{\alpha-\delta+\varepsilon}\Theta^{\beta-\gamma-\delta}H^{\gamma+\delta}L^{-\gamma-3\varepsilon}
\end{aligned}
\tag{4.4}
$$

where the 5 exponents α, β, γ, δ, and ε satisfy the 5 constraints

$$T : 1 - \gamma = 0, \tag{4.5a}$$

$$M : \alpha - \delta + \varepsilon = 0, \tag{4.5b}$$

$$\Theta : \beta - \gamma - \delta = 0, \tag{4.5c}$$

$$H : \gamma + \delta = 0, \tag{4.5d}$$

and

$$L : -\gamma - 3\varepsilon = 0. \tag{4.5e}$$

Their solution, $\alpha = -2/3$, $\beta = 0$, $\gamma = 1$, $\delta = -1$, and $\varepsilon = -1/3$, produces

$$\Delta t = C \cdot \frac{m^{2/3}\rho^{1/3}c_p}{k} \tag{4.6}$$

where C is a dimensionless number.

Note that the difference ΔT between the turkey's initial temperature and the oven temperature does not enter into this result (4.6). If we had instead of $\Delta T (= T_{oven} - T_o)$ separately included the turkey's initial temperature T_o, the oven temperature T_{oven}, and the turkey's desired interior temperature T_{inside} among the dimensional variables and constants, our analysis would have yielded

$$\Delta t = \frac{m^{2/3}\rho^{1/3}c_p}{k} \cdot g\left(\frac{T_o}{T_{oven}}, \frac{T_{inside}}{T_{oven}}\right) \tag{4.7}$$

where $g(x, y)$ is an undetermined function of two arguments.

Even so, (4.7) leaves out other possible variables – in particular those that determine shape. After all, a pancake-shaped "turkey," with relatively large surface area per volume, should cook more quickly than a spherical one. But suppose all turkeys are similarly shaped, their initial temperatures are identical, all ovens are set identically, and each turkey is cooked until its interior reaches that at which bacteria are killed (74 Celsius or 165 Fahrenheit). Then scaling (4.6), $\Delta t = C \cdot m^{2/3} \rho^{1/2} c_p / k$, suffices and only one datum is needed to determine the constant C.

It takes me about 4.25 hours to cook a 9.0-kilogram (20 pound) turkey in a 163 Celsius (325 Fahrenheit) oven. Since different turkeys have approximately the same thermal conductivity, specific heat, and mass density, these numbers produce the handy formula

$$\Delta t = 4.25 \cdot hours \cdot \left(\frac{m}{9.0 \cdot kg}\right)^{2/3}$$
$$= 1.0 \cdot hours \cdot \left(\frac{m}{kg}\right)^{2/3}. \tag{4.8}$$

While the proportionality constant in (4.8), based as it is on my experience, may not suit everyone, the scaling $\Delta t \propto m^{2/3}$ should apply to all cooks and to all turkeys.

The United States Department of Agriculture (USDA) publishes a table of cooking time versus turkey mass that seems to incorporate this scaling. Figure 4.1 shows the USDA data (translated into *SI* units) with error bars representing ranges given, for example, in the instruction "cook a 5.5–6.6 kilogram turkey for 3–3.75 hours." I have plotted $\Delta t = 1.0 \cdot m^{2/3}$ on a graph of the USDA data where m is in kilograms and Δt is in hours.

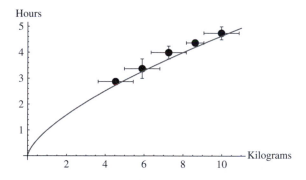

Figure 4.1. Data: USDA turkey cooking time recommendations. Solid line: $\Delta t = 1.0 \cdot m^{2/3}$.

4.5 Diffusion

Recall that a temperature gradient $\partial T / \partial x$ generates a heat flux q, that is, a rate of heat transfer per unit facing area, determined by (4.3), according to which

$$q = -k \frac{\partial T}{\partial x} \tag{4.9}$$

where k is the thermal conductivity of the material through which the heat conducts. However, (4.9) tells only half the story of heat conduction. For a non-uniform heat flux q, that is a non-vanishing $\partial q / \partial x$, inevitably leads to regions into which more heat enters than leaves. Furthermore, the temperature T of a region into which more heat enters than leaves increases at a rate $\partial T / \partial t$. Thus, we have

$$\rho c_p \frac{\partial T}{\partial t} = -\frac{\partial q}{\partial x} \tag{4.10}$$

where the negative sign indicates that if more heat enters a region than leaves, then $\partial q / \partial x < 0$ and the temperature increases, that is, $\partial T / \partial t > 0$. Combining (4.9) and (4.10) we find, providing the thermal conductivity k is uniform in space, that

$$\frac{\partial T}{\partial t} = \frac{k}{\rho c_p} \frac{\partial^2 T}{\partial x^2} \tag{4.11a}$$

or, equivalently,

$$\frac{\partial T}{\partial t} = D_T \frac{\partial^2 T}{\partial x^2} \tag{4.11b}$$

where

$$D_T = \frac{k}{\rho c_p}. \tag{4.12}$$

D_T is called the *thermal diffusivity*. The mechanism by which heat conducts through an object with constant thermal diffusivity D_T is called *diffusion*. Note that $[D_T] = T^{-1}L^2$.

The diffusion equation (4.11) is one of the basic partial differential equations of applied physics. Its single parameter D_T, the thermal diffusivity, determines the rate at which heat diffuses or conducts. When we are confident that a process is pure conduction, as in "Cooking a Turkey," we may construct our dimensional model with the thermal diffusivity D_T rather than with its separate components k, ρ, and c_p.

4.6 Making the Equations Dimensionless

We can also extract the scaling $\Delta t = C \cdot m^{2/3} \rho^{1/3} c_p / k$ that describes thermal diffusion directly from the diffusion equation (4.11b). Accordingly, we recast (4.11b), by multiplying and dividing equals by equals, into a form

$$\frac{\partial(T/T_o)}{\partial(t/\Delta t)} = \frac{D_T \Delta t}{l^2} \cdot \frac{\partial^2(T/T_o)}{\partial(x/l)^2} \tag{4.13}$$

where T_o, Δt, and l^2 are dimensional constants that characterize the initial temperature T_o and the time Δt required for a temperature T to penetrate a distance l. Therefore, the dimensionless product $D_T \Delta t / l^2$, which appears in the dimensionless equation (4.13), is equal to a dimensionless number C. Thus, given $\Delta t = C \cdot l^2 / D_T$ and $D_T = k/\rho c_p$, we find that $\Delta t = C \cdot l^2 \rho c_p / k$. Also since $l^2 \propto V^{2/3}$ and $V = m/\rho$ we find that $\Delta t = C' \cdot m^{2/3} \rho^{1/3} c_p / k$ – which reproduces (4.6).

Making the equations that govern a process or define a state dimensionless and then extracting the dimensionless product or products that parameterize these equations is a distinct method of dimensional analysis. According to this method we should list all the relevant equations and identify the dimensionless products entering into them. If there is only one dimensionless product, that product is equal to a dimensionless number. If there is more than one, the products are related to each other by an unknown function.

The process of making the equations dimensionless takes the place of the Rayleigh algorithm. Let's call this procedure *making the equations dimensionless* rather than the unattractive, if more common, *non-dimensionalization*. Of course, making the equations dimensionless implies that we know what those equations are. And sometimes we do not.

4.7 Example: Growth of Arctic Ice

For the purpose of being the first to reach the North Pole, the Norwegian explorer Fridtjof Nansen (1861–1930) allowed his ship, the *Fram* (in Norwegian "Forward"), to become frozen in the ice of the Laptev Sea (north of Siberia). Nansen believed that the natural drift of polar ice would carry the *Fram* across the North Pole. As it happened, Nansen made it only to 86 degrees north latitude (a record), but he and his men survived. Nansen later wrote an excellent account of his journey entitled *Farthest North*.

One of Nansen's occupations that first winter of his expedition (1884–1885) was to measure the thickness of the ice that surrounded the *Fram*. He noted that,

From measurements that were constantly being made, it appeared that the ice which
was formed in the autumn in October or November continued to increase in size
during the whole of the winter and out into the spring, but more slowly the thicker it
became. [18]

That the ice grew "more slowly the thicker it became" is confirmed by
dimensional analysis. Figure 4.2 illustrates the geometry.

The water below the layer of ice is close to the freezing temperature of seawater,
$-2.0\,\mathrm{C}\,(= T_o)$, while the temperature $T(< T_o)$ of the air above the layer of ice is,
during the Arctic winter, sub-freezing. Under these conditions the ice thickness λ
grows in time t. As the liquid water just under the layer of ice freezes, the heat
expelled from that liquid conducts through the ice up to the cooler air. Therefore,
this process, although one of heat transfer, is not one of pure diffusion but
also involves the fusion of liquid water into ice. The heat of fusion of water h,
the thermal conductivity of ice k, the mass density of ice ρ, and the mass-specific
heat capacity of ice at constant pressure c_p are all important. These symbols,
their descriptions, and their dimensional formulae are collected in Table 4.2.

Since there are 7 dimensional variables and constants and 5 imposed
dimensions, the analysis should produce 2 dimensionless products of the form
$\lambda^\alpha \Delta T^\beta k^\gamma h^\delta \rho^\varepsilon c_p^\phi$. Thus,

$$\left[\lambda t^\alpha \Delta T^\beta k^\gamma h^\delta \rho^\varepsilon c_p^\phi\right] = L T^\alpha \Theta^\beta \left(HT^{-1}L^{-1}\Theta^{-1}\right)^\gamma \left(HM^{-1}\right)^\delta \left(ML^{-3}\right)^\varepsilon \left(H\Theta^{-1}M^{-1}\right)^\phi$$
$$= L^{1-\gamma-3\varepsilon} T^{\alpha-\gamma} \Theta^{\beta-\gamma-\phi} H^{\gamma+\delta+\phi} M^{-\delta+\varepsilon-\phi}$$

$$(4.14)$$

Table 4.2

λ	Ice thickness	L
t	Time	T
$\Delta T \ (= T_o - T)$	Temperature difference	Θ
k	Thermal conductivity	$HT^{-1}L^{-1}\Theta^{-1}$
h	Heat of fusion	HM^{-1}
ρ	Mass density of water	ML^{-3}
c_p	Mass specific heat of ice	$H\Theta^{-1}M^{-1}$

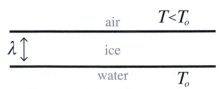

Figure 4.2. Layers of sub-freezing air, ice, and freezing water.

implies that

$$L : 1 - \gamma - 3\varepsilon = 0, \tag{4.15a}$$

$$T : \alpha - \gamma = 0, \tag{4.15b}$$

$$\Theta : \beta - \gamma - \phi = 0, \tag{4.15c}$$

$$H : \gamma + \delta + \phi = 0, \tag{4.15d}$$

and

$$M : -\delta + \varepsilon - \phi = 0. \tag{4.15e}$$

Constraints (4.15) are solved by $\alpha = \gamma = -1/2$, $\delta = -\beta$, $\varepsilon = 1/2$, and $\phi = \beta + 1/2$. Therefore, the dimensionless products are $\lambda(h\rho/t\Delta Tk)^{1/2}$ and $h/c_p\Delta T$, and the ice thickness λ is described by

$$\lambda = \sqrt{\frac{tk\Delta T}{h\rho}} \cdot f\left(\frac{h}{c_p\Delta T}\right) \tag{4.16}$$

where $f(x)$ is an undetermined function. Since the ice thickness λ increases as $t^{1/2}$, the sea ice surrounding the *Fram*, indeed, grew "more slowly the thicker it became."

If we had replaced the imposed dimension heat H with its energy equivalent ML^2T-2, an extraneous dimensionless product $th^2\rho/k\Delta T$ would have been produced – one that represents what we do not intend our model to allow: the dissipation of work into or the production of work out of internal energy.

Josef Stefan (1835–1893) was the first in 1891 to construct and solve a complete model of the ice-thickening process [19]. Since Stefan's time a mathematical problem in which a solution is required in a region whose boundary moves, as does the lower boundary of the ice in Figure 4.2, is called a *Stefan problem*.

Stefan's analysis was probably motivated by news generated from earlier Arctic explorations including that of the ill-fated Franklin expedition of 1845 in which all perished searching for the Northwest Passage, and the disastrous USS Jeanette expedition of 1879–1882 whose captain, George Washington De Long, and many of his crew died while searching for the mythical "open polar sea." Nansen's relatively successful expedition occupied three years, 1893–1896. Later (1910–1912), the *Fram* carried Roald Amundsen to Antarctica for his trek to the South Pole. To this day the Norwegian people lovingly preserve the *Fram* in an Oslo museum.

4.8 Example: Stack Effect

I once worked in a fine old building constructed in the 1920s. Its design included small passageways or chimneys that connected its upper floors to roof vents. By encouraging a natural flow of air – as illustrated in Figure 4.3 – these chimneys kept the upper floors of the building from becoming too hot in the summer.

Air of density ρ and mass-specific heat capacity c_p enters the room through an open window and flows out through a chimney of height h. If the inside temperature T_H becomes higher than the outside temperature $T_C (< T_H)$, a buoyant force lifts the air up the chimney at speed v – a tendency called the *stack effect*. How does v depend on the dimensional variables and constants whose dimensional formulae are collected in Table 4.3? Note that we do not include the heat H among the imposed dimensions. After all, the buoyant air does work on its environment as it expands and ascends the chimney.

From this list we may immediately eliminate the mass density ρ of the air since its dimensional formula ML^{-3} cannot be combined with the others in a dimensionless product. This leaves 6 dimensional variables and constants, v, g, h, c_p, $T_H - T_C$, and T_H, and 3 imposed dimensions, L, T, and Θ. Therefore, the

Table 4.3

v	Speed of air flow	LT^{-1}
g	Acceleration of gravity	LT^{-2}
h	Height of chimney	L
c_p	Mass specific heat	$L^2 T^{-2} \Theta^{-1}$
$T_H - T_C$	Temperature increment	Θ
T_H	Inside temperature	Θ
ρ	Mass density	ML^{-3}

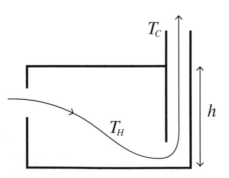

Figure 4.3. Natural convection through a chimney.

analysis should produce 3 dimensionless products. An inspection of the table reveals these to be $v/\sqrt{gh}$, $c_p(T_H - T_C)/gh$, and $(T_H - T_C)/T_H$. Therefore, the speed v of the airflow may be expressed as

$$v = \sqrt{gh} \cdot f\left[\frac{c_p(T_H - T_C)}{gh}, \frac{(T_H - T_C)}{T_H}\right]. \tag{4.17}$$

Dimensional analysis takes us no further.

Note, however, that the quantities gh and $c_p(T_H - T_C)$ are energies per unit mass and the ratio $(T_H - T_C)/T_H$, that is, $1 - T_C/T_H$, is the Carnot efficiency of an ideal heat engine operating between the two temperatures T_H and T_C. The stack effect, essentially, sets up a heat engine that generates work from a temperature difference. Because an engine whose Carnot efficiency vanishes produces no work, we know that $f(0, 0) = 0$. For this reason, the leading order term in an asymptotic expansion of $f(x, y)$ in terms of small x and y is given by $f(x, y) = C \cdot x^n y^m$ where n and m do not both vanish. Another kind of analysis shows that $n = 0$ and $m = 1/2$ in which case (4.17) becomes

$$v = C \cdot \sqrt{2gh}\left(1 - \frac{T_C}{T_H}\right) \tag{4.18}$$

where C is an undetermined dimensionless number.

Essential Ideas

The dimensional variable temperature T and the two dimensions Θ, denominating the dimension temperature, and H, denominating the imposed dimension heat, are introduced. *Heat* is "that energy transferred when no work is done." When no work is done on or by the system, as in the examples "Cooking a turkey" and "Growth of Arctic ice," the imposed dimension heat H increases the number of dimensions, decreases the number of dimensionless products generated, and produces a more informed result. When work is done, as in the example "Stack effect," the dimension H must be replaced by its energy equivalent ML^2T^{-2}.

Problems

4.1 **Ideal gas**. The ideal gas pressure equation of state can be expressed as $pV = nRT$ where p is the gas pressure, n is the number of moles in volume V, R is the "ideal gas constant," and T is the absolute temperature. Find the dimensional formula of R, that is, find $[R]$.

4.2 **Pure conduction**. A periodic change of temperature of frequency ω is impressed on the face of a half-infinite solid of diffusivity D_T. A wave of temperature variation conducts into the half-infinite solid. Find (a) how the wavelength λ of this variation and (b) its speed of propagation v depend on ω and D_R. [20]

4.3 **Conduction and convection**. A fluid of speed v flowing around a hot object conducts and convects heat away from the object. The temperature difference between the hot object and the cooler fluid is ΔT. The size of the object is l where $[l] = L$, the mass-specific heat capacity of the fluid is c_p, the mass density of the fluid is ρ, and the thermal conductivity of the fluid is k. Determine how the heat flux q depends on these dimensional variables and constants, that is, on v, ΔT, l, c_p, ρ, and k. Since this is a heat transfer problem, the heat H is an effective dimension along with M, L, T, and Θ. [21]

5

Electrodynamics and Plasma Physics

5.1 Maxwell's Equations

An electric charge density ρ generates an electric field $\underline{E}$ while moving charges compose a current density $\underline{J}$ that generates a magnetic field $\underline{B}$. James Clerk Maxwell (1831–1879) devised equations that relate these fields, $\underline{E}$ and $\underline{B}$, to their sources, ρ and $\underline{J}$. These are, in differential vector form,

$$\nabla \cdot \underline{E} = \frac{\rho}{\varepsilon_o}, \tag{5.1a}$$

$$\nabla \times \underline{E} = -\frac{\partial}{\partial t}\underline{B}, \tag{5.1b}$$

$$\nabla \cdot \underline{B} = 0, \tag{5.1c}$$

and

$$\nabla \times \underline{B} = \mu_o \underline{J} + \mu_o \varepsilon_o \frac{\partial}{\partial t}\underline{E}. \tag{5.1d}$$

The dimensional constant ε_o is known as the *vacuum permittivity* or, more prosaically, the *electric constant*, and μ_o as the *vacuum permeability* or the *magnetic constant*. Once created, electric $\underline{E}$ and magnetic $\underline{B}$ fields exert a force $\underline{F}$ on a charge q as described by the *Lorentz force law*

$$\underline{F} = q\left(\underline{E} + \underline{v} \times \underline{B}\right) \tag{5.2}$$

where $\underline{v}$ is the velocity of the charge. Equations (5.1) and (5.2) govern all electrodynamic phenomena.[a]

---

[a] Maxwell's equations can be cast into a form, favored by theorists, in terms of only one dimensional constant: the speed of light c. This is possible because the vacuum permeability μ_o is, in *SI* units, a conventional number $4\pi \cdot 10^{-7}$ and, for this reason, can be absorbed into the definition of the electric field and the charge density.

Maxwell first composed his eponymous equations (5.1) in 1865 from contributions made during the previous 80 years. He was especially indebted to Michael Faraday (1791–1867) and urged those who would understand his own work first to read Faraday's 1,100-page *Experimental Researches in Electricity*. Faraday pioneered the concept of *field* as opposed to the *action at a distance* view that prevailed in his time. According to *action at a distance*, charged particles exert forces directly on each other. Faraday believed, rather, that fields are real objects that mediate the interaction of charged particles. Maxwell's effort to mathematize Faraday's pictorial understanding of electric and magnetic fields led him to formulate equations (5.1). The second term on the right-hand side of the Ampere-Maxwell law (5.1d) introduces Maxwell's own contribution: the *displacement current* term $\mu_o \varepsilon_o (\partial E / \partial t)$.

Maxwell showed that the Ampere-Maxwell law (5.1d) and Faraday's law (5.1b) together allow for self-supporting electromagnetic field structures that detach from their sources and propagate through space at the speed of light. In this way Maxwell predicted that light waves are part of the spectrum of electromagnetic waves – a prediction Heinrich Hertz (1857–1897) verified experimentally, if accidentally, in 1886.

Maxwell's equations (5.1) require one new dimension, the electric charge, whose *SI* unit is the *coulomb* and whose symbol is Q. Therefore, the dimension of charge density ρ is QL^{-3} and of current density J is $QT^{-1}L^{-2}$. From $[\rho] = QL^{-3}$, $[J] = QT^{-1}L^{-2}$, and (5.2) we find that

$$[E] = MLT^{-2}Q^{-1}, \tag{5.3a}$$

and

$$[B] = MT^{-1}Q^{-1}. \tag{5.3b}$$

From Gauss's law (5.1a) and from (5.3a) we find that

$$[\varepsilon_o] = M^{-1}L^{-3}T^2Q^2 \tag{5.4a}$$

and from the Ampere-Maxwell law (5.1d) and from (5.3b) we find that

$$[\mu_o] = MLQ^{-2}. \tag{5.4b}$$

Note that since $[\mu_o \varepsilon_o] = L^{-2}T^2$, the dimensional constant $1/\sqrt{\mu_o \varepsilon_o}$ has the dimensional formula of speed LT^{-1}. Furthermore, the magnitude of $1/\sqrt{\mu_o \varepsilon_o}$, to three places $3.00 \cdot 10^8 \cdot m/s$, is the speed of light *in vacuo*. We will also make use of the potential difference or voltage V defined by $E = -\nabla V$. Therefore, given (5.3a), $[V] = ML^2T^{-2}Q^{-1}$.

A useful approximation of Maxwell's equations is one in which both Maxwell's displacement current term in (5.1d) and the right-hand side of

Faraday's law (5.1b) are unimportant. In this approximation, electric and magnetic fields decouple and each is generated separately from its own kind of source, electric fields from charges and magnetic fields from currents. This decoupling creates the separate realms of *electrostatics* and *magnetostatics*. The vacuum permittivity ε_o enters into the first and the vacuum permeability μ_o into the second.

5.2 Example: Oscillations of a Compass Needle

By the end of the eighteenth century, small magnetized pieces of iron had been used for six or seven centuries as compass needles to help guide seafarers. Also by the end of the eighteenth century, the direction of magnetic north relative to celestial north (*magnetic declination*) and the deflection of a compass needle in the vertical direction (*magnetic dip*) had been mapped over much of the Atlantic. Magnetic declination and dip together describe the direction of what was eventually called the *geomagnetic field*. But how can the magnitude of this field be measured?

In 1776, Jean Charles Borda (1733–1799), a French naval engineer who would shortly be commanding French warships aiding the American revolutionaries, had a clever idea. Just as the oscillation frequency ω of a pendulum is directly proportional to the square root of the magnitude g of the local acceleration of gravity so that $\omega \propto \sqrt{g}$, the oscillation frequency ω of a compass needle should be proportional to the square root of the magnitude B of the local geomagnetic field so that $\omega \propto \sqrt{B}$. By measuring the oscillation frequency ω of a compass needle Borda hoped to measure B. [22]

Because $[\omega] = T^{-1}$ and $[B] = MT^{-1}Q^{-1}$ the frequency ω must depend on more than the magnitude of B. Certainly ω should also depend on how strongly the compass needle responds to a magnetic field and on the compass needle's rotational inertia. The magnetic dipole moment μ of the compass needle (whose symbol should not be confused with that of the vacuum permeability μ_o) parameterizes the first and the needle's moment of inertia I the second. These symbols, their descriptions, and their dimensional formulae are found in Table 5.1.

Table 5.1

ω	Frequency	T^{-1}
B	Magnetic field strength	$MT^{-1}Q^{-1}$
μ	Magnetic dipole moment	$QT^{-1}L^2$
I	Moment of inertia	ML^2

Since there are 4 dimensional variables and constants and 4 imposed dimensions, these should not, according to the rule of thumb $N_P = N_V - N_D$, form a dimensionless product – unless the number of effective dimensions is less than 4. In the latter case the dimensionless product will have the form $\omega B^{\varepsilon} \mu^{\beta} I^{\gamma}$ where the exponents make

$$
\begin{aligned}
\left[\omega B^{\alpha} \mu^{\beta} I^{\gamma}\right] &= T^{-1}\left(MT^{-1}Q^{-1}\right)^{\alpha}\left(QT^{-1}L^{2}\right)^{\beta}\left(ML^{2}\right)^{\gamma} \\
&= T^{-1-\alpha-\beta}M^{\alpha+\gamma}Q^{-\alpha+\beta}L^{2\beta+2\gamma}
\end{aligned}
\tag{5.5}
$$

equal to 1. Therefore,

$$
T : -1 - \alpha - \beta = 0,
\tag{5.6a}
$$

$$
M : \alpha + \gamma = 0,
\tag{5.6b}
$$

$$
Q : -\alpha + \beta = 0,
\tag{5.6c}
$$

and

$$
L : 2\beta + 2\gamma = 0.
\tag{5.6d}
$$

Equations (5.6) are solved by $\alpha = -1/2, \beta = -1/2$, and $\gamma = 1/2$. In this way, we find that

$$
\omega = C \cdot \sqrt{\frac{B\mu}{I}}
\tag{5.7}
$$

where C is an undetermined dimensionless number.

Note that only two of the three equations (5.6b), (5.6c), and (5.6d) are linearly independent – a sure sign that only 3 of the 4 imposed dimensions are effective ones. A brief inspection of the table reveals that the three dimensions M, L, Q appear only in two combinations: ML^2 and QL^2T^{-1}. These latter are, respectively, dimensional formulae of the moment of inertia and the magnetic dipole moment. Therefore, there are only 3 effective dimensions, T, ML^2, and QL^2T^{-1}, rather than 4, M, T, L, and Q.

Scaling (5.7) confirms Borda's speculation that $\omega \propto \sqrt{B}$. In principle, Borda could have compared the magnitude B of the magnetic field at different places by measuring the oscillation frequency ω at different places with the same compass or with identically constructed compasses. Indeed, the French government commissioned a two-ship expedition in 1785, under the direction of the Comte de La Perouse, in part to measure the magnetic declination, dip, and magnitude at various places in the Pacific. Unfortunately, in 1788, the expedition's ships were wrecked on coral reefs off the coast of Vanikoro, one of the Solomon Islands. Evidence suggests that some of La Perouse's sailors survived for some years, even for some decades. A French search and rescue mission sighted Vanikoro in 1793 but, not realizing the expedition had foundered there, failed to land. In 1827, an Irish captain, Peter Dillon, commanding a British vessel, landed on Vanikoro where he bought "a ship's bell, a plank with a *fleur-de-lys* and guns with the maker's identification mark still visible" – all items of European manufacture, the remains of La Perouse's ships – but found no survivors. [23]

5.3 Example: Radiation from an Accelerating Charge

A charged particle moving with constant velocity is attended by electric and magnetic fields that move along with it. These fields do not detach from and propagate away from their source, that is, they do not compose electromagnetic waves. Only accelerating charges radiate electromagnetic waves that carry energy, momentum, and, when designed to do so, information away from their source. Cell phone and radio transmission towers, for instance, support antennas along whose arms electrons move back and forth. Because these electrons continually change their speed and direction of motion, they continually accelerate. And because they continually accelerate, they continually radiate electromagnetic waves.

Consider a charged particle with constant acceleration a, say, one moving with constant speed v in a circle of radius r so that its centripetal acceleration $a = v^2/r$. At what rate P does this charged particle radiate energy? Since radiation is an electromagnetic phenomenon, the power radiated P must depend on both of the dimensional constants that enter into Maxwell's equations, ε_o and μ_o, as well as on the charge q and its acceleration a. These dimensional variables and constants, their descriptions, and their dimensional formulae are collected in Table 5.2.

Since there are 5 dimensional variables and constants and 4 imposed dimensions, we expect 1 dimensionless product of the form $Pq^\alpha a^\beta \varepsilon_o^\gamma \mu_o^\delta$. We choose the exponents in order to make

$$\left[Pq^\alpha a^\beta \varepsilon_o^\gamma \mu_o^\delta \right] = ML^2 T^{-3} Q^\alpha \left(LT^{-2} \right)^\beta \left(M^{-1} L^{-3} T^2 Q^2 \right)^\gamma \left(MLQ^{-2} \right)^\delta$$
$$= M^{1-\gamma+\delta} L^{2+\beta-3\gamma+\delta} T^{-3-2\beta+2\gamma} Q^{\alpha+2\gamma-2\delta} \qquad (5.8)$$

Table 5.2

P	Power radiated	$ML^2 T^{-3}$
q	Charge	Q
a	Acceleration	LT^{-2}
ε_o	Vacuum permittivity	$M^{-1} L^{-3} T^2 Q^2$
μ_o	Vacuum permeability	MLQ^{-2}

equal to 1. Therefore,

$$M : 1 - \gamma + \delta = 0, \tag{5.9a}$$

$$L : 2 + \beta - 3\gamma + \delta = 0, \tag{5.9b}$$

$$T : -3 - 2\beta + 2\gamma = 0, \tag{5.9c}$$

and

$$Q : \alpha + 2\gamma - 2\delta = 0. \tag{5.9d}$$

The solution of (5.9) is: $\alpha = -2$, $\beta = -2$, $\gamma = -1/2$, and $\delta = -3/2$. Therefore, $Pq^{-2}a^{-2}\varepsilon_o^{-1/2}\mu_o^{-3/2}$ is dimensionless and so

$$P = C \cdot \frac{q^2 a2}{\varepsilon_o c^3} \tag{5.10}$$

where C is an undetermined dimensionless number. Note that in (5.10) we have replaced μ_o with c via $c = 1/\sqrt{\varepsilon_o \mu_o}$. J. J. Larmor first derived (5.10) with $C = (6\pi)^{-1}$ in 1897.

Larmor's result indicates that the planetary model of the hydrogen atom, for a short time in vogue after Rutherford's 1911 discovery of the atomic nucleus, is untenable. For the electron, in this case, would continually radiate energy and, therefore, continually descend into the nucleus. Niels Bohr was aware of this problem when, in 1913, he devised quantum conditions that not only saved the stability of the hydrogen atom but also predicted its emission and absorption spectra.

5.4 Example: Child's Law

The terms *anode* and *cathode* denote two parts of a *vacuum tube*, the simplest geometry of which is illustrated in Figure 5.1. Two parallel, metallic plates with surface area A are separated by a distance $s(\ll \sqrt{A})$. One of the metallic plates is charged negative (the cathode) and the other positive (the anode) with a potential difference ΔV between them. A flow of electrons of current density J is emitted from the cathode and accelerated toward the anode.

C. D. Child (1868–1933) discovered, in 1911, a relationship, now bearing his name, that determines the maximum current density J_{max} an electron tube with this geometry may produce. Because like charges repel, the larger the current density J, the less accelerating force on an electron just emitted from the cathode. Indeed, if J is large enough, an electron just emitted from the cathode will be so repelled by the electrons in the space between the two plates, it will experience no acceleration. This maximum current density J_{max} is called the *space charge limited current density*.

Table 5.3

J_{max}	Maximum current density	$QT^{-1}L^{-2}$
s	Plate separation	L
ΔV	Potential difference	$ML^2T^{-2}Q^{-1}$
ε_o	Vacuum permittivity	$Q^2L^{-3}M^{-1}T^{-2}$
e	Electron charge	Q
m_e	Electron mass	M

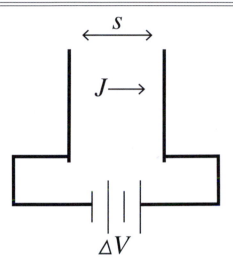

Figure 5.1. Parallel-plate geometry of an electron tube.

The space charge limited current density J_{max} is a function of the plate separation s, the potential difference ΔV, the vacuum permittivity ε_o, the electron charge e, and the electron mass m_e. These symbols, their descriptions, and their dimensional formulae are collected in Table 5.3.

The dimensionless products produced by these dimensional variables and constants assume a form $J_{max}s^\alpha \Delta V^\beta \varepsilon_o^\gamma e^\delta m_e^\varepsilon$ where the exponents are constrained by the requirement that

$$\left[J_{max}s^\alpha \Delta V^\beta \varepsilon_o^\gamma e^\delta m_e^\varepsilon\right] = QT^{-1}L^{-2}L^\alpha \left(ML^2T^{-2}Q^{-1}\right)^\beta \left(Q^2L^{-3}M^{-1}T^{-2}\right)^\gamma Q^\delta M^\varepsilon$$
$$= Q^{1-\beta+2\gamma+\delta}T^{-1-2\beta-2\gamma}L^{-2+\alpha+2\beta-3\gamma}M^{\beta-\gamma+\varepsilon}$$

$$(5.11)$$

be equal to 1. Thus,

$$C : 1 - \beta + 2\gamma + \delta, \qquad (5.12a)$$
$$T : -1 - 2\beta - 2\gamma = 0, \qquad (5.12b)$$
$$L : -2 + \alpha + 2\beta - 3\gamma = 0, \qquad (5.12c)$$

and

$$M : \beta - \gamma + \varepsilon = 0. \tag{5.12d}$$

These constraints are solved by $\beta = -7/2 + \alpha$, $\gamma = -3 + \alpha$, $\delta = 3/2 - \alpha$, and $\varepsilon_o = 1/2$. The two dimensionless products produced, $J_{max} e^{3/2} m_e^{1/2} / [\Delta V^{7/2} \varepsilon_o^3]$ and $s\Delta V\varepsilon_o/e$, are related by

$$J_{max} = \frac{\varepsilon_o^3 \Delta V^{7/2}}{e^{3/2} m_e^{1/2}} \cdot f\left[\frac{e}{s\Delta V\varepsilon_o}\right] \tag{5.13}$$

where $f(x)$ is an undetermined function. This is as far as dimensional analysis takes us.

However, if one could be assured, as is here the case, that the electron charge e and mass m_e enter into (5.13) only in the charge to mass ratio e/m_e, then it must be that $f(x) = C \cdot x^2$ and so

$$J_{max} = C \cdot \frac{\varepsilon_o \Delta V^{3/2}}{s^2} \sqrt{\frac{e}{m_e}}. \tag{5.14}$$

A more detailed analysis reveals that this is so and that $C = 4\sqrt{2}/9$. The cloud of electrons in front of the cathode that maintains the space charge limited current is called a *virtual cathode*.

5.5 Plasma Physics

A plasma is a collection of disassociated electrons and positive ions. Powerful electrostatic forces usually keep plasmas globally charge neutral. Even so, plasmas can support waves that create local regions of net charge and current density and, therefore, local regions of non-vanishing electric and magnetic fields. Most natural and artificially produced plasmas are gases that have been to some degree ionized. Yet the free electrons and immobile ions within a metallic conductor and the charges within an electrolyte can also be considered plasmas.

Plasmas are produced naturally on the surface of the earth in lightning strikes and in the upper atmosphere or *ionosphere* by ionizing solar radiation. The arc of a welder's torch is plasma, as is the hottest part of a flame. Plasmas are used to etch integrated circuits. While our nearby environment is largely plasma-free, most of the matter of the universe is hot enough and tenuous enough to maintain significant ionization and is, therefore, in the plasma state. The sun, for instance, is plasma, as is most of the interstellar and intergalactic gas.

5.6 Example: Plasma Oscillations

Consider a plasma composed of electrons of density n_e and positive ions of density n_i and charge state Z such that $Zn_i = n_e$. If the plasma electrons are shifted from a position of equilibrium, they will be pulled back and accelerate toward their equilibrium position, overshoot that position because of their inertia, and continue to oscillate at a frequency ω_p called the plasma frequency. Since the electrons are much less massive than the ions, most of this oscillatory motion will be in the electrons.

How does the plasma frequency ω_p depend on the electron density n_e, the electron charge e, the electron mass m_e, and the vacuum permittivity ε_o? These 5 dimensional variables and constants are denominated in 4 imposed dimensions M, L, T, and Q. Therefore, according to the rule of thumb $N_P = N_V - N_D$, the analysis should produce 1 dimensionless product. These symbols, their descriptions, and their dimensional formulae are collected in Table 5.4.

In this case we can easily construct the dimensionless product without employing the Rayleigh algorithm. One starts with ε_o and divides or multiplies by powers of the other quantities in each case so that the division or multiplication eliminates one of the imposed dimensions M, L, T, or Q from the dimensional formula of the product. The single dimensionless product produced in this way is $\omega_p^2 m_e \varepsilon_o / n_e e^2$. Thus

$$\omega_p^2 = C \cdot \frac{n_e e^2}{m_e \varepsilon_o} \tag{5.15}$$

where C is a dimensionless number. As one might expect, the larger the electron mass m_e, the smaller the oscillation frequency ω_p. Also, the larger the restoring force as characterized by the parameter $n_e e^2 / \varepsilon_o$, the larger the oscillation frequency ω_p. According to a more detailed analysis, $C = 1$.

Table 5.4

ω_p	Plasma frequency	T^{-1}
n_e	Electron density	L^{-3}
e	Electron charge	Q
m_e	Electron mass	M
ε_o	Vacuum permittivity	$M^{-1}L^{-3}T^2Q^2$

5.7 Example: Pinch Effect

The electrostatic attraction of unlike charges and the electrostatic repulsion of like charges are not the only kinds of electromagnetic interaction. Whenever like charges, say electrons, all move or drift in the same direction, they compose a current whose different streams attract each other. Oppositely directed streams of current repel each other. These interactions, mediated as they are by a magnetic field, are normally much weaker than electrostatic attraction and repulsion. But when a dense cloud of electrons drifts through an equally dense cloud or array of relatively stationary ions, the electrostatic forces are neutralized and leave the magnetic force dominant. The resulting attraction of different streams of current is called the *pinch effect*.

When generated by the high current of a lightning bolt, the pinch effect may be so large as to implode a rigid metallic lightning rod carrying that current. In 1905, two Australian engineers, J. A. Pollock and S. Barraclough, examined the crushed and distorted remains of a lightning rod, originally a copper tube that had been erected above a refinery in the province of New South Wales (Australia). Their conclusion was that the copper tube had, indeed, been struck by lightning and imploded by the pinch effect.

In mid-twentieth century the pinch effect was thought capable of producing plasma hot enough and dense enough to support the nuclear fusion of light nuclei. The idea was that the initial part of a current pulse directed along a long straight wire would vaporize and ionize the material of the wire and the gas around it while the pinch effect associated with the remaining part of the pulse would compress and heat the plasma thus produced. Eventually, the outwardly directed pressure exerted by the plasma would balance the inwardly directed pinch force. The relation between the plasma current I, the plasma radius R, the plasma density per unit length N_L, and the plasma absolute temperature T when these forces are balanced can be found with dimensional analysis. Figure 5.2 illustrates the "z-pinch" geometry of this cylindrical plasma.

Because this is not a heat transfer problem but rather one in which the magnetic field compresses and heats the plasma, the dimension H is not independent of M, L, and T. Boltzmann's constant k_B and the vacuum

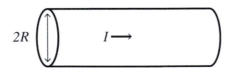

Figure 5.2. Z-Pinch Geometry.

Table 5.5

I	Current	QT^{-1}
R	Radius	L
N_L	Density per unit length	L^{-1}
T	Temperature	Θ
k_B	Boltzmann's constant	$ML^2T^{-2}\Theta^{-1}$
μ_o	Vacuum permittivity	MLQ^{-2}

permittivity μ_o should be included among the dimensional variables and constants. Furthermore, since we are concerned only with the fully pinched and balanced configuration, we need include neither a compression time nor the inertia of the plasma particles. (But see Problem 5.5.) Table 5.5 collects these variables and constants, their descriptions, and their dimensional formulae.

The table clearly indicates that Boltzmann's constant k_B must multiply the absolute temperature T – otherwise the dimension of temperature Θ would not be eliminated from the dimensionless product. For this reason we save ourselves a little effort by including these two dimensional variables and constants only in the product k_BT. Then we have 5 dimensionless variables and constants I, R, N_L, k_BT, and μ_o denominated in 4 imposed dimensions M, L, T, and Q.

The dimensionless products are of the form $IR^\alpha N_L^\beta (k_BT)^\gamma \mu_o^\delta$ where the exponents make this form dimensionless. Thus,

$$\left[IR^\alpha N_L^\beta (k_BT)^\gamma \mu_o^\delta \right] = QT^{-1}L^\alpha (L^{-1})^\beta (ML^2T^{-2})^\gamma (MLQ^{-2})^\delta$$
$$= Q^{1-2\delta} T^{-1-2\gamma} L^{\alpha-\beta+2\gamma+\delta} M^{\gamma+\delta} \tag{5.16}$$

implies that

$$Q : 1 - 2\delta = 0, \tag{5.17a}$$

$$T : -1 - 2\gamma = 0, \tag{5.17b}$$

$$L : \alpha - \beta + 2\gamma + \delta = 0, \tag{5.17c}$$

and

$$M : \gamma + \delta = 0. \tag{5.17d}$$

The solution of (5.17) is $\beta = \alpha - 1/2$, $\gamma = -1/2$, and $\delta = 1/2$. Note that only two of the three equations (5.17a), (5.17b), and (5.17d) are linearly independent. Evidently, only 3 dimensions are effective. An inspection of the table reveals these to be QT^{-1}, MT^{-2}, and L or alternatively and equivalently QT^{-1}, MLT^{-2}, and L. These latter identify the total current QT^{-1} and the force MLT^{-2}, that is, F, as effective dimensions. With enough foresight we could we have adopted these as imposed dimensions.

The dimensionless products assume the forms $I\sqrt{\mu_o/(N_L k_B T)}$ and RN_L. Therefore,

$$I^2 = \frac{N_L k_B T}{\mu_o} \cdot f(RN_L) \qquad (5.18)$$

where $f(x)$ is an undetermined function. This is as far as dimensional analysis takes us. In most applications $RN_L \gg 1$. A more detailed analysis indicates that as $RN_L \rightarrow \infty$, $f(RN_L) \rightarrow C$ where $C = 8\pi$.

Essential Ideas

Electrodynamics introduces three new dimensional fundamental constants: the vacuum permittivity ε_o, the vacuum permeability μ_o, and the speed of light c. Because ε_o, μ_o, and c are related by $c = 1/\sqrt{\varepsilon_o \mu_o}$, no more than two of these three are needed in any one description. One new dimension, the charge Q, is also introduced. To model an electrostatic state or process, one needs ε_o, and to model a magnetostatic state or process, one needs μ_o. Electromagnetic processes in which electric and magnetic fields are coupled require both ε_o and μ_o.

Problems

5.1 **Electret oscillations**. An *electret* is a material, possibly quartz or polymer, that can, for an extended period, hold a charge separation and in this way create a semi-permanent electric dipole moment p for which $[p] = QL$. Imagine a needle-like electret with moment of inertia I mounted so that it can freely rotate and align itself with an applied electric field of magnitude E. Use dimensional analysis to determine its frequency ω of oscillation around its equilibrium position. See Section 5.2.

5.2 **Fields**. A particle with charge q and constant velocity of magnitude v passes within a distance d of an observer. (a) How does the maximum electric field strength E detected by the observer depend on these variables and ε_o and μ_o? (b) How does the maximum magnetic field strength B detected by the observer depend on these variables and ε_o and μ_o?

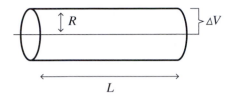

Figure 5.3. Geometry of a wire cathode and a concentric cylinder anode.

5.3 **Cylindrical Child's law**. The context is Section 5.4 on Child's law. Suppose the cathode is a long straight wire and the anode a concentric cylinder of radius R and length L as shown in Figure 5.3. A potential difference ΔV between the cathode and anode causes a maximum current I_{max} to flow between the two. Find an expression for the maximum current per unit length I_{max}/L between the cathode and anode that is parallel in structure to (5.13) and also to (5.14).

5.4 **Plasma oscillation**. The context is Section 5.6. Use the Rayleigh algorithm to show that only one dimensionless product $\omega_p \sqrt{m_e \varepsilon_o / n_e e^2}$ is produced from the dimensional variables and constants ω_p, m_e, ε_o, n_e, and e.

5.5 **Pinch effect compression time**. The context is Section 5.7. Use the Rayleigh algorithm to determine an expression for the time Δt required for a current I to cause a "cold," that is, a zero-pressure, plasma of initial radius R_o composed of ions of mass m and density per unit length N_L to become fully compressed. (Include the vacuum permeability μ_o among the dimensional variables and constants.)

5.6 **Pinch effect**. Again, the context is Section 5.7 on the pinch effect. Reformulate the table listing the symbols, names, and dimensional formula of 5 dimensional variables and constants, I, R, N_L, $k_B T$, and μ_o. In doing so express the dimensional formulae in terms of 3 effective dimensions: a current QT^{-1}, a pressure per unit length $ML^{-2}T^{-2}$, and a length L.

6

Quantum Physics

6.1 Planck's Constant

Max Planck's professional goal in the period 1894–1900 was to model the properties of electromagnetic radiation in thermal equilibrium with *blackbody* materials, so called because they absorb (and emit) electromagnetic radiation at all frequencies. Thus, the radiation in equilibrium with a blackbody is called *blackbody radiation*. At first, Planck (1858–1947) made little progress – classical descriptions of blackbody radiation failed to reproduce even its most basic properties. And before 1900, classical physics was the whole of physics.

Planck eventually looked for a simple postulate that could mathematically reproduce the properties of blackbody radiation even if that postulate went beyond what was then known. What he discovered in 1900 was the bold assertion that a blackbody can absorb and emit radiation of frequency v only in countable bundles or *quanta* each with energy E proportional to v. Thus, $E = hv$ where h is a constant now called *Planck's constant*. Planck's postulate led to expressions that fit the available data well when h assumed a value close to

$$h = 6.63 \cdot 10^{-34} \cdot kg \cdot m^2/s \qquad (6.1)$$

where (6.1) gives, to three places, the current value of Planck's constant in *SI* units. Note that $[h] = ML^2T^{-1}$.

While Planck's postulate was a great success, it took some time for physicists to fathom its consequences. In 1905, Albert Einstein (1879–1955) showed that Planck's quanta, later called *photons*, also explained the photoelectric effect, while Arthur Holly Compton, in 1922, showed that a photon and an electron preserve their total momentum and energy as they interact with one another. Eventually, it became clear that Planck had discovered a new realm of physics – the realm of the quantum. Today we recognize Planck's constant h as a sure sign of quantum physics.

6.2 Example: Blackbody Radiation

The equation of state of radiation in a cavity surrounded by blackbody material can be explored dimensionally. Because blackbody radiation is a quantum phenomenon, its energy density E/V depends on Planck's constant h as well as on the speed of light c and on the product $k_B T$ where k_B is Boltzmann's constant and T is the absolute temperature of the radiation. These symbols, their descriptions, and their dimensional formulae are collected in Table 6.1.

Because there are 4 dimensional variables and constants, E/V, h, c, and $k_B T$, and 3 imposed dimensions, M, L, and T, the dimensional analysis should, according to the rule of thumb $N_P = N_V - N_D$, produce 1 dimensionless product. We find this product by making $(E/V)h^\alpha (k_B T)^\beta c^\gamma$ dimensionless. Thus,

$$\left[(E/V)h^\alpha (k_B T)^\beta c^\gamma \right] = \left(ML^{-1}T^{-2} \right) \left(ML^2 T^{-1} \right)^\alpha \left(ML^2 T^{-2} \right)^\beta \left(LT^{-1} \right)^\gamma \tag{6.2}$$
$$= M^{1+\alpha+\beta} L^{-1+2\alpha+2\beta} T^{-2-\alpha-2\beta-\gamma}$$

implies that

$$M : 1 + \alpha + \beta = 0, \tag{6.3a}$$

$$L : -1 + 2\alpha + 2\beta + \gamma = 0, \tag{6.3b}$$

and

$$T : -2 - \alpha - 2\beta - \gamma = 0. \tag{6.3c}$$

Equations (6.3) are solved by $\alpha = 3$, $\beta = -4$, and $\gamma = 3$. Therefore, the product $(E/V)(hc)^3 (k_B T)^{-4}$ is dimensionless and

$$\frac{E}{V} = C \cdot \frac{(k_B T)^4}{(hc)^3} \tag{6.4}$$

where C is a dimensionless number. Equation (6.4) with $C = 8\pi^5 / 15$ is known as the *Stefan-Boltzmann law*.

Table 6.1

E/V	Energy density	$ML^{-1}T^{-2}$
h	Planck's constant	$ML^2 T^{-1}$
$k_B T$	Boltzmann's constant times absolute temperature	$ML^2 T^{-2}$
c	Speed of light	LT^{-1}

6.3 Example: Spectral Energy Density of Blackbody Radiation

The energy density E/V of blackbody radiation is composed of electromagnetic waves of frequency v associated with photons having energy hv whose density is non-uniform over the spectrum of electromagnetic waves from $v = 0$ to $v = \infty$. Thus, we may ask, "How does the energy density of blackbody radiation per differential frequency, that is, $V^{-1}(dE/dv)$, depend upon frequency v?" This quantity $V^{-1}(dE/dv)$, known as the *spectral energy density* of blackbody radiation, depends on Planck's constant h, on the normalized absolute temperature $k_B T$, and on the speed of light c. These symbols, their descriptions, and their dimensional formulae are collected in Table 6.2.

Since there are 5 dimensional variables and constants and 3 imposed dimensions, we should find 2 dimensionless products. These assume a form $(dE/dv)V^{-1}v^\alpha(k_B T)^\beta h^\gamma c^\delta$ that the exponents α, β, γ, and δ render dimensionless. Thus

$$\left[(dE/dv)V^{-1}v^\alpha(k_B T)^\beta h^\gamma c^\delta\right] = ML^{-1}T^{-1}\left(T^{-1}\right)^\alpha\left(ML^2T^{-2}\right)^\beta\left(ML^2T^{-1}\right)^\gamma\left(LT^{-1}\right)^\delta$$
$$= M^{1+\beta+\gamma}L^{-1+2\beta+2\gamma+\delta}T^{-1-\alpha-2\beta-\gamma-\delta}$$

$$(6.5)$$

implies

$$M : 1 + \beta + \gamma = 0, \tag{6.6a}$$
$$L : -1 + 2\beta + 2\gamma + \delta = 0, \tag{6.6b}$$

and

$$T : -1 - \alpha - 2\beta - \gamma - \delta = 0. \tag{6.6c}$$

These are solved by $\beta = -3 - \alpha$, $\gamma = 2 + \alpha$, $\Delta = 3$. Therefore, $(dE/dv)V^{-1}h^2c^3(k_B T)^{-3}$ and $hv/k_B T$ are dimensionless and so

Table 6.2

$(dE/dv)V^{-1}$	Spectral energy density	$ML^{-1}T^{-1}$
v	Frequency	T^{-1}
$k_B T$	Boltzmann's constant times absolute temperature	ML^2T^{-2}
h	Planck's constant	ML^2T^{-1}
c	Speed of light	LT^{-1}

$$V^{-1}\left(\frac{dE}{dv}\right) = \frac{(k_B T)^3}{h^2 c^3} \cdot f\left(\frac{hv}{k_B T}\right) \tag{6.7}$$

where the function $f(x)$ is undetermined by dimensional analysis.

However, the form of the undetermined function $f(x)$ is constrained by the requirement that the integral of the spectral energy density $V^{-1}(dE/dv)$ over all frequencies is the average energy density E/V. Symbolically,

$$\int_0^\infty V^{-1}\left(\frac{dE}{dv}\right) dv = \left(\frac{E}{V}\right) \tag{6.8}$$

and so given (6.7),

$$\frac{(k_B T)^3}{h^2 c^3} \int_0^\infty f\left(\frac{hv}{k_B T}\right) dv = \frac{E}{V}. \tag{6.9}$$

Shifting the integration variable from v to x where $x = hv/k_B T$ transforms (6.9) into

$$\frac{(k_B T)^4}{(hc)^3} \int_0^\infty f(x) dx = \frac{E}{V}. \tag{6.10}$$

Given the equation of state (6.4) equation (6.10) becomes

$$\int_0^\infty f(x) dx = C. \tag{6.11}$$

This result relates the function $f(x)$ and the value of the number C, which otherwise remain undetermined. A more detailed analysis finds that $f(x) = x^3/(e^x - 1)$ and so, given (6.11), $C = 8\pi^5/15$.

6.4 Example: Bohr Model

Niels Bohr (1885–1962) used Planck's constant h to forge a new understanding of the hydrogen atom. According to the model Bohr proposed in 1913, the hydrogen atom is composed of an electron, with mass m_e and charge e, orbiting, in specially defined non-radiating orbits, a proton with mass $m_p (= 1836 m_e)$. Since the electron and proton are bound together electrostatically, the vacuum permittivity ε_o helps determine these orbits. Because Bohr's orbits are, by hypothesis, non-radiating and we assume the electron speed in a

Table 6.3

r_1	Bohr radius	L
m_e	Electron mass	M
e	Electron charge	Q
ε_o	Vacuum permittivity	$M^{-1}L^{-3}T^2Q^2$
h	Planck's constant	ML^2T^{-1}

hydrogen atom is non-relativistic, the speed of light c is not relevant. Therefore, the dimensional constants m_e, e, ε_o, and h complete the list of those that
describe Bohr's hydrogen atom.

Bohr's model determines, among other quantities, a *ground state* orbital
radius r_1 that characterizes the hydrogen atom's spatial extent. Dimensional
analysis determines how r_1 depends on m_e, e, ε_o, and h. Table 6.3 collects
these symbols, their descriptions, and their dimensional formulae.

Since this table contains 5 fundamental, dimensional constants denominated
in 4 imposed dimensions, there should be only 1 dimensionless product having
the form $r_1 m_e^\alpha e^\beta \varepsilon_o^\gamma h^\delta$. The exponents α, β, γ, and δ that determine this product
are those that make $r_1 m_e^\alpha e^\beta \varepsilon_o^\gamma h^\delta$ dimensionless. Therefore,

$$\begin{aligned}
\left[r_1 m_e^\alpha e^\beta \varepsilon_o^\gamma h^\delta \right] &= L M^\alpha Q^\beta \left(M^{-1}L^{-3}T^2Q^2 \right)^\gamma \left(ML^2T^{-1} \right)^\delta \\
&= L^{1-3\gamma+2\delta} M^{\alpha-\gamma+\delta} Q^{\beta+2\gamma} T^{2\gamma-\delta}
\end{aligned} \tag{6.12}$$

implies that

$$L : 1 - 3\gamma + 2\delta = 0, \tag{6.13a}$$

$$M : \alpha - \gamma + \delta = 0, \tag{6.13b}$$

$$Q : \beta + 2\gamma = 0, \tag{6.13c}$$

and

$$T : 2\gamma - \delta = 0. \tag{6.13d}$$

The solutions to equations (6.13) are $\alpha = 1$, $\beta = 2$, $\gamma = -1$, and $\delta = -2$.
Therefore, $r_1 m_e e^2 / \left(\varepsilon_o h^2 \right)$ is dimensionless and

$$r_1 = C \cdot \frac{\varepsilon_o h^2}{e^2 m_e} \tag{6.14}$$

where C is a dimensionless number. Bohr's own, simple analysis determines
that $C = 1/\pi$. The quantity r_1 with $C = 1/\pi$ in (6.14) is called the *Bohr radius*.

The above dimensional analysis is often done in a slightly different way, in
which the vacuum permittivity ε_o is replaced with $4\pi\varepsilon_o$ and Planck's constant h

with its "reduced value" $\hbar(= h/2\pi)$. Then the dimensionless product produced is $r_1 m_e e^2/(4\pi\varepsilon_o\hbar^2)$ and the resulting scaling is

$$r_1 = C' \cdot \frac{4\pi\varepsilon_o\hbar^2}{e^2 m_e} \tag{6.15}$$

with $C' = 1$. While this result suggests that we should use $4\pi\varepsilon_o$ in place of ε_o and use $\hbar$ in place of h, we have no reason to believe that this tactic will always produce proportionality constants equal to 1.

6.5 Atomic Units

The dimensional constants m_e, e, ε_o, and h enter into all models in which mechanics, electrostatics, and quantum physics play a role. Because atomic physics draws on these subjects, $\varepsilon_o h^2/e^2 m_e$ (proportional to the Bohr radius) characterizes lengths in atomic physics. Similarly, the time $\varepsilon_o^2 h^3/e^4 m_e$ (see Problem 6.2) and, of course, the mass m_e and charge e similarly characterize atomic physics.

The rule of thumb $N_P = N_V - N_D$ suggests and the Rayleigh algorithm confirms that only one length can be fashioned out of the 4 dimensional constants, m_e, e, ε_o, and h, whose dimensional formulae are expressed in terms of the 4 imposed dimensions, M, L, T, and Q. Similarly, only one time follows from these same 4 dimensional constants. In all, the 4 dimensional fundamental constants, m_e, e, ε_o, and h, produce 4 unique fundamental, characteristic quantities: a length $\varepsilon_o h^2/e^2 m_e$, a time $\varepsilon_o^2 h^3/e^4 m_e$, a mass m_e, and a charge e. These compose what we call a characterizing system of units or a *scale* – a scale of atomic-sized units. The elements of this scale, their formulae, and their values to three places in *SI* units are listed in Table 6.4.

Table 6.4

		atomic scale	
Mass	m_e	$9.11 \cdot 10^{-31}$ kg	Electron mass
Length	$\dfrac{\varepsilon_o h^2}{e^2 m_e}$	$1.67 \cdot 10^{-10}$ m	Bohr radius
Time	$\dfrac{\varepsilon_o^2 h^3}{e^4 m_e}$	$3.82 \cdot 10^{-17}$ s	Period of electron orbiting at Bohr radius
Charge	e	$1.60 \cdot 10^{-19}$ C	Electron charge

Because atomic physicists exert much effort calculating with m_e, e, ε_o, and h, they sometimes take the shortcut of setting each of these to unity, that is, so that $m_e = 1$, $e = 1$, $\varepsilon_o = 1$, and $h = 1$. The effect of this shortcut is that all masses produced in this way are in units of m_e. For instance, $m_p = 1836$ means $m_p = 1836 m_e$. Furthermore, all lengths are in units of $\varepsilon_o h^2 / e^2 m_e$, all times are in units of $\varepsilon_o^2 h^3 / e^4 m_e$, and all charges are in units of e.

6.6 Example: Quantum Ideal Gas

A gas is *ideal* if its particles do not interact with one another at a distance or through a field. We already know that the equation of state of a classical ideal gas $p = (N/V)k_B T$ relates its pressure p, its particle density N/V, and its absolute temperature T where k_B is Boltzmann's constant. How do quantum effects modify this equation of state?

Certainly Planck's constant h should enter into any quantum version of $p = (N/V)k_B T$. But the Rayleigh algorithm applied to the 4 dimensional variables and constants p, N/V, $k_B T$, and h that are denominated in terms of the 3 imposed dimensions M, L, and T generates exactly one dimensionless product $pV/(Nk_B T)$.

Possibly we have ignored some dimensional variable or constant that effectively brings Planck's constant h into play. Since, by definition, no force fields play a role, the dimensional constants G, ε_o, μ_o, and e are not involved. Because we limit our model particles to non-relativistic speeds, neither is the speed of light c. It seems that only the mass m of the ideal gas particles remains to be considered. In retrospect, it may surprise us that m does not enter directly into the classical equation of state $p = (N/V)k_B T$. (But review Section 1.7.)

Therefore, we allow the pressure p of a quantum ideal gas to depend on N/V, $k_B T$, h, and m. These 5 symbols, their descriptions, and their dimensional formulae are collected in Table 6.5.

Table 6.5

p	Pressure	$ML^{-1}T^{-2}$
N/V	Number density	L^{-3}
$k_B T$	Boltzmann's constant times absolute temperature	$ML^2 T^{-2}$
h	Planck's constant	$ML^2 T^{-1}$
m	Particle mass	M

Accordingly, $N_V = 5$ and $N_D = 3$. Therefore, we expect, according to the rule of thumb $N_P = N_V - N_D$, that the Rayleigh algorithm will produce 2 dimensionless products, one of which we already know is $pV/(Nk_BT)$. The dimensionless products formed are those that make $p(N/V)^\alpha (k_B T)^\beta h^\gamma m^\delta$ dimensionless. Given that

$$\left[p(N/V)^\alpha (k_B T)^\beta h^\gamma m^\delta \right] = (ML^{-1}T^{-2})L^{-3\alpha} (ML^2 T^{-2})^\beta (ML^2 T^{-1})^\gamma M^\delta$$
$$= M^{1+\beta+\gamma+\delta} L^{-1-3\alpha+2\beta+2\gamma} T^{-2-2\beta-\gamma},$$

$$(6.16)$$

we have

$$M : 1 + \beta + \gamma + \delta = 0, \qquad (6.17a)$$
$$L : -1 - 3\alpha + 2\beta + 2\gamma = 0, \qquad (6.17b)$$

and

$$T : -2 - 2\beta - \gamma = 0. \qquad (6.17c)$$

The solution to (6.17) is $\beta = -5/2 - 3\alpha/2$, $\gamma = 3 + 3\alpha$, and $\delta = -3/2 - 3\alpha/2$. Therefore, $ph^3 / \left[m^{3/2} (k_B T)^{5/2} \right]$ and $(N/V)h^3/(mk_B T)^{3/2}$ are the 2 dimensionless products. Multiplying the first by the inverse of the second produces $pV/(Nk_B T)$. Therefore, we may write

$$\frac{pV}{Nk_B T} = f \left[\frac{(N/V)h^3}{(mk_B T)^{3/2}} \right] \qquad (6.18)$$

where $f(x)$ is undetermined. This is as far as dimensional analysis takes us. However, we do know that $p = (N/V)k_B T$ must be recovered from (6.18) in the limit of vanishing Planck's constant h. Therefore, as $(N/V)h^3/(mk_B T)^{3/2} \to 0$, $f(0) \to 1$.

Quantum effects dominate in the high-density, low-temperature, small-mass regime for which $(N/V)h^3/(mk_B T)^{3/2} \gg 1$. If in this (quantum) regime the function $f(x)$ is replaced with the power law $f(x) \approx C \cdot x^n$, then (6.18) becomes

$$\frac{pV}{Nk_B T} = C \cdot \left[\frac{(N/V)h^3}{(mk_B T)^{3/2}} \right]^n \qquad (6.19)$$

where C and n are undetermined. Our physics sense takes us a little further. Because we expect that as $T \to 0$, the pressure p should not diverge, it must be that $n \leq 2/3$.

In the special case $n = 2/3$ (6.19) becomes

$$p = C \cdot \left(\frac{N}{V} \right)^{5/3} \left(\frac{h^2}{m} \right). \qquad (6.20)$$

This result (6.20) with $C = (3/\pi)^{2/3}/20$ describes the *degeneracy pressure* exerted by a gas of particles that observe Fermi-Dirac statistics according to which no two particles in the same region can occupy the same single-particle quantum state. Equation (6.20) also expresses the dependence of the critical pressure at which a gas of *bosons* begins to transition to a condensed state. Bosons are particles that observe Bose-Einstein statistics according to which any number of particles may occupy one single-particle quantum state.

Of course, dimensional analysis does not answer or even pose crucial questions such as "Why is there a degeneracy pressure?" or "Why does a Bose-Einstein condensate exist?" Only a more complete theory can frame, much less answer, these questions.

6.7 Example: Quantized Radiation from an Accelerating Charge

According to the result of Section 5.4, the rate P at which a charge q moving with speed v in a circle of radius r radiates electromagnetic waves that carry away energy is given by

$$P = C \cdot \frac{q^2 a^2}{\varepsilon_o c^3} \tag{6.21}$$

where $a = v^2/r$ and C is an undetermined dimensionless number. But we now know that this electromagnetic energy is emitted in quanta or photons each with energy $h\nu$ for various frequencies ν. How does this knowledge modify (6.21)?

Clearly, Planck's constant h must enter into any quantum modification of (6.21). Therefore, this quantum modification must be a relation among the following dimensional variables and constants: the power radiated P, the charge q, its acceleration a, the vacuum permittivity ε_o, the speed of light c, and Planck's constant h. These symbols, descriptive names, and dimensional formulae are listed in Table 6.6.

Table 6.6

P	Power radiated	ML^2T^{-3}
q	Charge	Q
a	Acceleration	LT^{-2}
ε_o	Vacuum permittivity	$M^{-1}L^{-3}T^2Q^2$
c	Speed of light	LT^{-1}
h	Planck's constant	ML^2T^{-1}

Since there are 6 dimensional variables and constants and 4 imposed dimensions, we expect 2 dimensionless product of the form $Pq^\alpha a^\beta \varepsilon_o^\gamma c^\delta h^\varepsilon$. We have already identified, in Section 5.4, one of these: $P\varepsilon_o c^3/(q^2 a^2)$. The other we find in the usual way, that is, by choosing exponents α, β, γ, δ, and ε that render the form $Pq^\alpha a^\beta \varepsilon_o^\gamma c^\delta h^\varepsilon$ dimensionless. Therefore,

$$\left[Pq^\alpha a^\beta \varepsilon_o^\gamma c^\delta h^\varepsilon\right] = ML^2 T^{-3} Q^\alpha \left(LT^{-2}\right)^\beta \left(M^{-1} L^{-3} T^2 Q^2\right)^\gamma \left(LT^{-1}\right)^\delta \left(ML^2 T^{-1}\right)^\varepsilon$$
$$= M^{1-\gamma+\varepsilon} L^{2+\beta-3\gamma+\delta+2\varepsilon} T^{-3-2\beta+2\gamma-\delta-\varepsilon} Q^{\alpha+2\gamma}$$

$$(6.22)$$

implies that

$$M : 1 - \gamma + \varepsilon = 0, \qquad\qquad (6.23a)$$

$$L : 2 + \beta - 3\gamma + \delta + 2\varepsilon = 0, \qquad\qquad (6.23b)$$

$$T : -3 - 2\beta + 2\gamma - \delta - \varepsilon = 0, \qquad\qquad (6.23c)$$

and

$$Q : \alpha + 2\gamma = 0. \qquad\qquad (6.23d)$$

The solution of (6.23) is: $\alpha = -2$, $\beta = -2$, $\gamma = -\alpha/2$, $\delta = 2 - \alpha/2$, and $\varepsilon = -1 - \alpha/2$. In this way we find the two dimensionless products to be $Pc^2/a^2 h$ and $q/\sqrt{\varepsilon_o ch}$. By multiplying the first by the inverse square of the second, we eliminate h and produce, as expected, $P\varepsilon_o c^3/(q^2 a^2)$. The square of the second independent, dimensionless product is $q^2/\varepsilon_o ch$. Thus, we may write

$$P = \frac{q^2 a2}{\varepsilon_o c^3} \cdot f\left(\frac{q^2}{\varepsilon_o ch}\right) \qquad\qquad (6.24)$$

where the function $f(x)$ remains undetermined.

The dimensionless product $q^2/(2\varepsilon_o hc)$ with the electron charge e replacing q is called the *fine structure constant* – of which more will be said in Chapter 7. The undetermined function in (6.24) is not easily discovered or expressed. One approach to determining $f(x)$ is to expand the semi-classical (and thus semi-quantum) expression for P in terms of powers of the fine structure constant $e^2/(2\varepsilon_o ch)[\approx 0.007]$. [24]

Essential Ideas

The dimensional model of a quantized state or process must include the dimensional constant known as Planck's constant h.

Problems

6.1 **Ground state energy**. The context is Section 6.4. Determine how the ground state energy E_1 of a hydrogen atom depends on relevant dimensional constants.

6.2 **Atomic time unit**. (a) Use the Rayleigh algorithm to show that only one quantity with the dimension of time, $\varepsilon_o^2 h^3 / e^4 m_e$, is fashioned out of the fundamental constants m_e, e, ε_o, and h. (b) Show that this time is equal to Planck's constant divided by the energy found in Problem 6.1.

6.3 **Ideal gas**. Show that, as claimed in Section 6.6, the Rayleigh algorithm applied to the dimensional variables and constants P, N/V, $k_B T$, and h generates one dimensionless product $PV/(Nk_B T)$.

7

Dimensional Cosmology

7.1 The Fundamental Dimensional Constants

Each subject we have explored thus far, mechanics and the mechanics of fluids, thermal physics, electrodynamics, and quantum physics, has introduced one or more fundamental dimensional constants: the gravitational constant G, Boltzmann's constant k_B, the vacuum permittivity ε_o, the speed of light c, the electron charge e, the electron mass m_e, and Planck's constant h. These subjects, considered expansively, encompass most of what we know of the physical world, and these fundamental constants quantify that knowledge. These seven constants, their descriptive names and symbols, their numerical values to three places in *SI* units, and their dimensional formulae are collected in Table 7.1.[a]

What I find remarkable is that these constants are related to one another through their dimensional formulae in ways that compose a *dimensional structure* or *cosmology*. Here I use the word *cosmology* broadly to denote the general structure of the universe, its elements, and its laws. This chapter seeks to uncover that cosmology. The results, partial and preliminary though they may be, are those that can be obtained with the methods of dimensional analysis. Other methods may take us further, but dimensional analysis brings us close to the boundary separating the known from the unknown.

7.2 Eddington-Dirac Number

We simplify our initial investigation in two ways, first by eliminating Boltzmann's constant k_B from consideration and second by concentrating on

---

[a] I have included the vacuum permittivity ε_o and the speed of light c in place of the electromagnetic constants ε_o and μ_o. Given that $c = 1/\sqrt{\varepsilon_o\mu_o}$ only two of the three constants ε_o, μ_o, and c are independent. Therefore, only two are needed in any one description.

Table 7.1

Name	Symbol	Value in SI units	Dimensional Formula
Gravitational constant	G	$6.67 \cdot 10^{-11}$	$M^{-1}L^3T^{-2}$
Boltzmann's constant	k_B	$1.38 \cdot 10^{-23}$	$ML^2T^{-2}\Theta^{-1}$
Vacuum permittivity	ε_o	$8.85 \cdot 10^{-12}$	$M^{-1}L^{-3}T^2Q^2$
Speed of light	c	$3.00 \cdot 10^8$	LT^{-1}
Electron charge	e	$1.60 \cdot 10^{-19}$	Q
Electron mass	m_e	$9.11 \cdot 10^{-31}$	M
Planck's constant	h	$6.63 \cdot 10^{-34}$	ML^2T^{-1}

the dimensional structure of classical physics. A brief inspection of the above table reveals that the dimension of temperature Θ appears only in Boltzmann's constant k_B. Consequently, k_B cannot enter into a dimensionless product with the other constants – and finding these dimensionless products is our first task.

Of course, Boltzmann's constant k_B has its uses. Besides converting the dimension of temperature Θ into that of energy ML^2T^{-2} and *vice versa*, Boltzmann's constant k_B provides entropy with its dimensional formula. After all, the statistical entropy S of an isolated system is given by $S = k_B\ln\Omega$ where Ω is the number of microstates accessible to that system. Therefore, $[S] = [k_B] = ML^2T^{-2}\Theta^{-1}$.

Once we eliminate Boltzmann's constant k_B and the non-classical Planck's constant h from the list of fundamental constants, the remaining 5 classical ones, G, ε_0, c, e, and m_e, characterize gravitational, electrostatic, and electromagnetic interactions, and the properties of an electron. Because these 5 constants are denominated in 4 imposed dimensions, M, L, T, and Q, they produce among themselves, according to the rule of thumb, 1 dimensionless product. The Rayleigh algorithm easily produces this dimensionless product: $e^2/\left(\varepsilon_o Gm_e^2\right)$.

Two electrons separated by a distance r illumine its meaning. These two electrons repel each other with an electrostatic force $e^2/\left(4\pi\varepsilon_o r^2\right)$ and attract each other with a gravitational force Gm_e^2/r^2. Therefore, the ratio $e^2/\left(4\pi\varepsilon_o Gm_e^2\right)$, or more simply $e^2/\left(\varepsilon_o Gm_e^2\right)$, characterizes the relative strength of these two forces.

The value of the ratio,

$$\frac{e^2}{4\pi\varepsilon_o Gm_e^2} = 4.16 \cdot 10^{42}, \tag{7.1}$$

fascinated Arthur Eddington (1882–1944) and Paul Dirac (1902–1984). Eddington noticed that 10^{42} is approximately the square root of the number of atoms in the observable universe, while Dirac found that 10^{42} is approximately the age of the universe in units of a classical unit of time

$e^2/(\varepsilon_o m_e c^3) \approx 10^{-23}$s. Eddington and Dirac used these correlations as the basis of speculations that, while interesting, have failed to bear fruit. Even so, we honor their efforts by calling $e^2/(4\pi\varepsilon_o G m_e^2)$ the *Eddington-Dirac ratio*.

The large size of the Eddington-Dirac ratio $e^2/(4\pi\varepsilon_o G m_e^2)$ means that the electrostatic force is much stronger than the gravitational force. Gravity dominates the large-scale structure of the universe only because large numbers of positive and negative charges tend to neutralize each other. Electrostatics dominates the small-scale structure of the universe. Together these two forces make possible a universe with an amazing variety of structures from atoms, to living organisms, to galaxies.

One consequence of the Eddington-Dirac ratio is the dimensional equivalence

$$[e^2/\varepsilon_o] = [Gm_e^2] \tag{7.2}$$

or $[e^2] = [\varepsilon_o G m_e^2]$ or $[\varepsilon_o] = [e^2/Gm_e^2]$. We may, for instance, substitute e^2/ε_o for Gm_e^2 in any term and preserve that term's dimensional formula. Of course, in doing so we shift its numerical value by a factor of 10^{42}.

7.3 Example: Gravitational-Electrostatic Mass

What quantity with the dimension of mass can be composed out of the three fundamental constants e, ε_o, and G? The dimensional equivalence (7.2), that is, $[e^2/\varepsilon_o] = [Gm_e^2]$, contains the answer. For if $[e^2/\varepsilon_o] = [Gm_e^2]$, then $[e^2/\varepsilon_o] = [G][m_e^2]$, $[e^2/\varepsilon_o] = [G][m_e]^2$, and so $[m_e] = [e/\sqrt{G\varepsilon_o}]$. Therefore, the quantity $e/\sqrt{G\varepsilon_o}$ has the dimension of mass – the mass of a spherical particle with charge e whose gravitational self-attraction is large enough to balance its electrostatic self-repulsion. I call $e/\sqrt{G\varepsilon_o}$ the *gravitational-electrostatic mass*.

7.4 Example: Classical Electron Radius

What quantity having the dimension of *length* can be composed out of the fundamental constants m_e, e, ε_o, and c? Since the Eddington-Dirac ratio does not contain the speed of light c, this question cannot be answered by using the dimensional equivalence (7.2) based on that dimensionless product.

Instead, denote the length sought with l. Given that the 5 dimensional quantities, l, m_e, e, ε_o, and c, are expressed in terms of 4 imposed dimensions,

Table 7.2

l	Characteristic length	L
m_e	Electron mass	M
e	Electron charge	Q
ε_o	Permittivity of free space	$M^{-1}L^{-1}T^2Q^2$
c	Speed of light	LT^{-1}

M, L, T, and Q, as shown in Table 7.2, the Raleigh algorithm applied to these quantities should produce 1 dimensionless product.

In this case, we do not need the Rayleigh algorithm to uncover the dimensionless product. We simply multiply and divide ε_o by powers of the other constants until the result becomes dimensionless. The dimensionless product produced in this way is $lm_e\varepsilon_o c^2/e^2$. Therefore, $e^2/(\varepsilon_o m_e c^2)$ has the dimension of length. This length is, apart from a dimensionless factor, the radius r of a spherical electron whose potential electrostatic energy, roughly $e^2/(4\pi\varepsilon_o r)$, is equal to its rest energy $m_e c^2$. The length $e^2/(4\pi\varepsilon_o m_e c^2)$ produced in this way is equal to $2.81 \cdot 10^{-15} \cdot m$ and is sometimes called the *classical electron radius*.

7.5 Classical Scales

From among the five classical, fundamental dimensional constants G, ε_o, c, e, and m_e we chose the following four ε_o, c, e, and m_e to form a quantity $e^2/(\varepsilon_o m_e c^2)$, the classical electron length, with the dimension of length. We can also divide this length $e^2/(\varepsilon_o m_e c^2)$ by the speed of light c to form a time $e^2/(\varepsilon_o m_e c^3)$. The Rayleigh algorithm indicates that these are the only quantities with the dimensions of length and time that are combinations of ε_o, c, e, and m_e.

This length and this time along with the electron mass m_e and charge e form a set of four quantities with dimensions M, L, T, and Q. It is customary to call this set of quantities a *scale*, in this case, the *classical scale*.[b,c]

---

[b] The name *classical scale* is something of a misnomer, since, as we shall soon see, there are five scales composed solely of classical, fundamental, dimensional constants.

[c] One could always add to the classical or any other scale consisting of a mass, length, and time, a characteristic temperature defined by a quantity with the dimensional formula of energy ML^2T^{-2} divided by Boltzmann's constant k_B.

Table 7.3

		classical scale	
Mass	m_e	$9.11 \cdot 10^{-31}$ kg	Electron mass
Length	$\dfrac{e^2}{\varepsilon_o m_e c^2}$	$3.53 \cdot 10^{-14}$ m	Classical electron radius
Time	$\dfrac{e^2}{\varepsilon_o m_e c^3}$	$1.18 \cdot 10^{-22}$ s	Time for light to cross a classical electron radius
Charge	e	$1.60 \cdot 10^{-19}$ C	Electron charge

The numerical values of these quantities, to three places in *SI* units, are collected in Table 7.3.

Other classical masses, lengths, durations, and charges can be formed in similar fashion, that is, by applying the Rayleigh algorithm to other sets of 4 dimensional constants taken from the 5 classical ones: G, ε_o, c, e, and m_e. There are in all $5!/4!1!$ or 5 such classical scales. These are collected in the first five rows of Table 7.4 in Section 7.7.

7.6 Fine Structure Constant

Restoring Planck's constant h to the classical, dimensional constants (and again excluding Boltzmann's constant k_B) composes a set of 6 fundamental, dimensional constants: G, ε_o, c, e, m_e, and h. Since their dimensional formulae are expressed in terms of 4 imposed dimensions, M, L, T, and Q, they can be combined into 2 dimensionless products. One of these, we know, is proportional to the Eddington-Dirac ratio $e^2/(4\pi\varepsilon_o G m_e^2)$. The other, which we met for the first time in Section 6.7, is $e^2/(\varepsilon_o h c)$. The particular form $e^2/(2\varepsilon_o h c)$ is called the *fine structure constant*.

Arnold Sommerfeld (1868–1951) observed that the ratio of the velocity of an electron in the first Bohr orbit to the speed of light is the fine structure constant. But the fine structure constant also carries with it another meaning. Because it is proportional to the ratio of the potential energy of two electrons separated by a distance d, that is, $e^2/(4\pi\varepsilon_o d)$, to the energy of a photon with wavelength d, that is, hc/d, the fine structure constant regulates the interaction of electrons and photons. Indeed, $e^2/(2\varepsilon_o h c)$ characterizes a generalization of Maxwell's equations, called *quantum electrodynamics* or *QED*, in which electrons and photons interact.

The value of the fine structure constant was first thought to be exactly $1/137$. Consequently, a number of physicists, including Wolfgang Pauli (1900–1958), sought the meaning of this apparently special integer 137. As it happened, Pauli was unsuccessful in this effort and balefully pointed out the number of his hospital room, 137, to a visitor shortly before his death. We now know that the fine structure constant is only approximately $1/137$. The current value of its inverse is, to 12 places, 137.035999139. Nevertheless, since the equation

$$\frac{e^2}{2\varepsilon_o hc} = \frac{1}{137} = 0.00730 \tag{7.5}$$

is accurate to three places, the number 137 is a handy way to remember the value of the fine structure constant.

No one has been able to derive numerical values of either the Eddington-Dirac ratio or the fine structure constant. Instead, these dimensionless products or their constituent dimensional constants must be measured. Yet there are physicists who believe the task of physics will not be complete until the value of all the fundamental dimensional constants can be derived from some very deep principle – now unknown to us.

The fine structure constant $e^2/(2\varepsilon_o hc)$ provides a way of generating dimensional quantities with the dimensions of M, L, T, and Q that contain Planck's constant h from quantities that do not. One simply uses dimensional equivalences like

$$[e^2/\varepsilon_o] = [hc] \tag{7.6}$$

or $[e^2] = [\varepsilon_o hc]$ or $[c] = [e^2/h\varepsilon_o]$ to transform classical quantities into quantum ones that contain Planck's constant h. See, for example, Section 7.9.

7.7 The 15 Scales

From the set of 6 fundamental dimensional constants, G, ε_o, c, e, m_e, and h, 4 can be taken, without regard to order, in $6!/4!2!$ or 15 different ways. Employing the Rayleigh algorithm on each of these 15 sets of 4 constants, we derive quantities with the dimensions of M, L, T, and Q.

Of course, the Rayleigh algorithm produces a particular quantity only when that quantity exists. And occasionally it does not. For instance, because the dimensional formulae of the 4 constants G, m_e, c, and h do not contain the dimension Q, the Rayleigh algorithm cannot generate, from these

Table 7.4

Constants	Mass (kilograms)	Length (meters)	Time (seconds)	Charge (coulombs)
G, m_e, e, ε_o	$m_e \sim 10^{-30}$ $\sqrt{e^2/\varepsilon_o G} \sim 10^{-8}$			$e \sim 10^{-19}$ $\sqrt{Gm_e^2\varepsilon_o} \sim 10^{-41}$
G, m_e, ε_o, c	$m_e \sim 10^{-30}$	$\dfrac{Gm_e}{c^2} \sim 10^{-57}$	$\dfrac{Gm_e}{c^3} \sim 10^{-66}$	$\sqrt{Gm_e^2\varepsilon_o} \sim 10^{-41}$
G, m_e, e, c	$m_e \sim 10^{-30}$	$\dfrac{Gm_e}{c^2} \sim 10^{-57}$	$\dfrac{Gm_e}{c^3} \sim 10^{-66}$	$e \sim 10^{-19}$
G, e, ε_o, c	$\sqrt{\dfrac{e^2}{\varepsilon_o G}} \sim 10^{-8}$	$\sqrt{\dfrac{Ge^2}{\varepsilon_o c^4}} \sim 10^{-36}$	$\sqrt{\dfrac{Ge^2}{\varepsilon_o c^6}} \sim 10^{-44}$	$e \sim 10^{-19}$
m_e, e, ε_o, c **classical scale**	$m_e \sim 10^{-30}$	$\dfrac{e^2}{\varepsilon_o m_e c^2} \sim 10^{-14}$	$\dfrac{e^2}{\varepsilon_o m_e c^3} \sim 10^{-22}$	$e \sim 10^{-19}$
G, m_e, e, h	$m_e \sim 10^{-30}$	$\dfrac{h^2}{Gm_e^3} \sim 10^{34}$	$\dfrac{h^3}{G^2 m_e^5} \sim 10^{71}$	$e \sim 10^{-19}$
G, m_e, ε_o, h	$m_e \sim 10^{-30}$	$\dfrac{h^2}{Gm_e^3} \sim 10^{34}$	$\dfrac{h^3}{G^2 m_e^5} \sim 10^{71}$	$\sqrt{Gm_e^2\varepsilon_o} \sim 10^{-41}$
G, m_e, c, h	$m_e \sim 10^{-30}$ $\sqrt{\dfrac{ch}{G}} \sim 10^{-7}$	$\dfrac{h}{m_e c} \sim 10^{-12}$	$\dfrac{h}{m_e c^2} \sim 10^{-20}$	
G, e, ε_o, h	$\sqrt{\dfrac{e^2}{\varepsilon_o G}} \sim 10^{-8}$	$\dfrac{\varepsilon_o^{3/2} h^2 G^{1/2}}{e^3} \sim 10^{-32}$	$\dfrac{\varepsilon_o^{5/2} h^3 G^{1/2}}{e^5} \sim 10^{-38}$	$e \sim 10^{-19}$
m_e, e, ε_o, h **atomic scale**	$m_e \sim 10^{-30}$	$\dfrac{\varepsilon_o h^2}{e^2 m_e} \sim 10^{-10}$	$\dfrac{\varepsilon_o^2 h^3}{e^4 m_e} \sim 10^{-17}$	$e \sim 10^{-19}$
G, e, c, h	$\sqrt{\dfrac{ch}{G}} \sim 10^{-7}$	$\sqrt{\dfrac{Gh}{c^3}} \sim 10^{-35}$	$\sqrt{\dfrac{Gh}{c^5}} \sim 10^{-43}$	$e \sim 10^{-19}$
m_e, e, c, h	$m_e \sim 10^{-30}$	$\dfrac{h}{m_e c} \sim 10^{-12}$	$\dfrac{h}{m_e c^2} \sim 10^{-20}$	$e \sim 10^{-19}$
G, ε_o, c, h **Planck scale**	$\sqrt{\dfrac{ch}{G}} \sim 10^{-7}$	$\sqrt{\dfrac{Gh}{c^3}} \sim 10^{-35}$	$\sqrt{\dfrac{Gh}{c^5}} \sim 10^{-43}$	$\sqrt{\varepsilon_o ch} \sim 10^{-18}$
m_e, ε_o, c, h	$m_e \sim 10^{-30}$	$\dfrac{h}{m_e c} \sim 10^{-12}$	$\dfrac{h}{m_e c^2} \sim 10^{-20}$	$\sqrt{\varepsilon_o ch} \sim 10^{-18}$
e, ε_o, c, h				$e \sim 10^{-19}$ $\sqrt{\varepsilon_o ch} \sim 10^{-18}$

constants, a quantity with the dimension Q. Therefore, this particular set of 4 possible dimensional quantities remains incomplete.

Each of the 15 sets of dimensional quantities derived from 4 fundamental dimensional constants composes a scale. The 15 scales, classical and quantum, complete and incomplete, are listed in Table 7.4. Three of them have been given special names: the *classical scale*, based on m_e, e, ε_o, and c, the *atomic scale* based on m_e, e, ε_o, and h, and the *Planck scale* based on G, ε_o, c, and h. Attending each dimensional combination is its value in *SI* units to the nearest order of magnitude.[d,e,f]

7.8 Planck Scale

Max Planck (1858–1947) was quite taken with the "naturalness" of what later were called *Planck scale units*. [25][g] Unlike SI or Gaussian units or other units in widespread use, Planck units do not depend on the properties of arbitrarily chosen, if universally recognized, measures like the kilogram, nor do they depend on the properties of any one particle. The dimensional formulae and the values of quantities that make up the Planck scale are listed in Table 7.5. The Planck scale is the only one of the 15 possible ones in which Planck's constant h appears in all four characteristic quantities.

Table 7.5

	Planck scale	
Mass	$\sqrt{\dfrac{ch}{G}}$	$2.18 \cdot 10^{-8}$ kg
Length	$\sqrt{\dfrac{Gh}{c^3}}$	$1.62 \cdot 10^{-35}$ m
Time	$\sqrt{\dfrac{Gh}{c^5}}$	$5.39 \cdot 10^{-44}$ s
Charge	$\sqrt{\varepsilon_o ch}$	$1.33 \cdot 10^{-18}$ C

---

[d] The quantities in the table proportional to the classical electron radius, the classical duration, and the Planck mass, length, time, and charge omit factors of 4π and 2 that are included in their customary definitions.

[e] Of course, from any one complete scale of characterizing quantities many other quantities can be derived by composition. For instance, given a characterizing mass, length, time, and charge, one can derive a characterizing speed, momentum, kinetic energy, angular momentum, current, magnetic moment, and so on.

[f] Note that of these 15 scales 3 are incomplete. The fundamental constants foundational to each of these 3 incomplete scales combine, in turn, to make the Eddington-Dirac number $e^2/(\varepsilon_o Gm_e^2)$ from G, m_e, e, and ε_o, the fine structure constant $e^2/(\varepsilon_o hc)$ from e, ε_o, c, and h, and their quotient ch/Gm_e^2 from G, m_e, c, and h. The existence of each of these dimensionless products reduces the number of characterizing quantities comprising each of the incomplete scales.

[g] Planck's constant h is often replaced by the reduced Planck's constant $\hbar (= h/2\pi)$ and ε_o with $4\pi\varepsilon_o$ in definitions of the Planck mass, length, time, and charge.

Planck scale quantities are those at which gravitational forces and quantum effects interact and, consequently, can be interpreted in these terms. For instance, the Planck length $\sqrt{Gh/c^3}$ is the geometrical mean of the Schwarzchild radius mG/c^2 and the Compton length h/mc of an object with mass m. The Schwarzchild radius mG/c^2 is that radius of a sphere of mass m whose escape velocity becomes the speed of light c, while the Compton wavelength h/mc is the de Broglie wavelength of an object with momentum mc where here and elsewhere m refers to an arbitrary rest mass. The Planck time $\sqrt{Gh/c^3}$ is that required for light to cross a Planck distance. Furthermore, the Planck mass $\sqrt{ch/G}$ is that mass whose Compton wavelength h/mc is equal to the Planck length $\sqrt{Gh/c^3}$. And the Planck charge $\sqrt{\varepsilon_o ch}$ is that of oppositely charged particles separated by distance d equal to the wavelength of a photon with energy, hc/d, sufficient to disassociate the particles, that is, one for which $hc/d = e^2/\varepsilon_o d$.

7.9 Example: Planck Mass

The Planck mass $\sqrt{hc/G}$ is easily derived from the electron mass m_e by using the dimensional equivalences built into the Eddington-Dirac ratio $e^2/\left(4\pi\varepsilon_o G m_e^2\right)$ and the fine structure constant $e^2/(2\varepsilon_o hc)$. In particular, we use the equivalence $[m_e] = \left[\sqrt{e^2/\varepsilon_o G}\right]$, derived from the Eddington-Dirac ratio, to identify a mass $\sqrt{e^2/\varepsilon_o G}$. Then we use the equivalence $[e^2/\varepsilon_0] = [hc]$, derived from the fine structure constant, to generate the Planck mass $\sqrt{hc/G}$. Interestingly, the Planck mass $\left(\sim 10^{-8} \cdot kg\right)$ is relatively large on a human scale considering that the Planck length $\left(\sim 10^{-35} \cdot m\right)$ and Planck time $\left(\sim 10^{-43} \cdot s\right)$ are so small. The best digital balances can resolve masses to within a Planck mass.

7.10 The 22 Quantities

The $60(= 15 \times 4)$ positions in the table of Section 7.6 are filled with 58 quantities. Of these 22 are unique. Each of these 22 unique quantities can be interpreted in terms of certain kinds of interactions (gravitational, electrostatic, and electromagnetic), certain effects (quantum mechanical), and certain particle properties (electron charge and mass).

The 22 quantities separated into categories with the dimensions of M, L, T, and Q create sequences of magnitudes. The masses in kilograms, from largest to smallest, are

$$10^{-7}, 10^{-8}, \text{and } 10^{-30}.$$

The lengths in meters, from largest to smallest, are

$$10^{34}, 10^{-10}, 10^{-12}, 10^{-14}, 10^{-32}, 10^{-35}, 10^{-36}, \text{and } 10^{-57}.$$

The durations in seconds, from largest to smallest, are

$$10^{77}, 10^{-17}, 10^{-20}, 10^{-22}, 10^{-38}, 10^{-43}, 10^{-44}, \text{and } 10^{-66}.$$

And, finally, the charges in coulombs, from largest to smallest, are

$$10^{-18}, 10^{-19}, \text{and } 10^{-41}.$$

One is drawn to especially large and small magnitudes. For instance, the charge $\sqrt{Gm_e^2\varepsilon_o} \left(\sim 10^{-41} \cdot C\right)$ is a factor of 10^{21} times smaller than the charge of an electron. This charge $\sqrt{Gm_e^2\varepsilon_o}$ is that of a spherical particle of mass m_e whose repulsive self-electrostatic force and attractive self-gravitational force are equal – a *gravitational-electrostatic charge*.

The length $h^2/Gm_e^3 \left(\sim 10^{34} \cdot m\right)$ is a factor of 10^8 times larger than the size of the observable universe $\left(\sim 10^{26} \cdot m\right)$. This distance is the separation of two uncharged masses m_e each pulled by gravity from rest at "infinity" until equal to their de Broglie wavelengths. In contrast, the very small length $Gm_e/c^2 \left(\sim 10^{-57} \cdot m\right)$ is the Schwarzchild radius of an object with the mass of an electron.

Even more startling is the duration $h^3/G^2m_e^5 \left(\sim 10^{71} \cdot s\right)$ – at least a factor of 10^{56} times longer than the age of the universe $\left(\sim 4 \cdot 10^{17} \cdot s\right)$. This duration $h^3/G^2m_e^5$ is, according to the Heisenberg uncertainty principle, the quantum lifetime of a gravitating sphere of mass m_e and radius h^2/Gm_e^3. By "quantum lifetime" $h^3/G^2m_e^5$ I mean the maximum lifetime Δt of an object having energy $\Delta E \left(= G^2m_e^5/h^2\right)$ that may, according to the Heisenberg uncertainty principle $\Delta t\Delta E \leq h$, fluctuate into existence.

Other quantities are less fancifully interpreted. For instance, $h/m_ec \left(\sim 10^{-12} \cdot m\right)$ is the Compton wavelength of an electron, and the quantity $h/m_ec^2 \left(\sim 10^{-20} \cdot s\right)$ is the time required for light to traverse this wavelength.

We have in this chapter uncovered dimensionless products and dimensional scales and quantities using only the methods of dimensional analysis. As we have come to expect, dimensional analysis takes us only so far and no further. Consequently, these dimensionless products and dimensional scales and quantities may seem more the input than the substance of a complete dimensional cosmology.

Essential Ideas

The 6 fundamental constants G, ε_o, c, e, m_e, and h generate two dimensionless products, the Eddington-Dirac number $e^2/(4\pi\varepsilon_o Gm_e^2)$ and the fine structure constant $e^2/(2\varepsilon_o hc)$. These 6 constants also generate quantities with dimensions of mass, length, time, and charge that compose scales at which various forces (gravity and electromagnetism), quantum effects, and the properties of fundamental particle interact.

Problems

7.1 **Eddington-Dirac number**. (a) Derive the dimensionless product $e^2/(\varepsilon_0 Gm_e^2)$ by applying the Rayleigh algorithm to the set of fundamental, classical, dimensional constants: G, ε_0, μ_0, e, and m_e. (b) Check to see that $e^2/(\varepsilon_0 Gm_e^2)$ is dimensionless by using the dimensional formulae in Table 7.1 of Section 7.1.

7.2 **Fine structure constant**. Show that the fine structure constant $e^2/(2\varepsilon_0 hc)$ is dimensionless by using the dimensional formulae in Table 7.1 of Section 7.1.

7.3 **Planck scale**. Derive the Planck length, time, and charge from the classical electron length $h/m_e c$, time $h/m_e c^2$, and charge e by substituting dimensional equivalences built into the Eddington-Dirac number $e^2/(4\pi\varepsilon_0 Gm_e^2)$ and the fine structure constant $e^2/(2hc)$ as illustrated in Section 7.9.

7.4 **Electron scales? Speed of light scale?** According to Section 7.8, "The Planck scale is the only one of the 15 in which Planck's constant h appears in all four characteristic quantities." But there other scales in which a particular fundamental constant appears in all four characteristic quantities.
   (a) Identify two scales that could be called an "electron mass scale" because m_e appears in all four characteristic quantities.
   (b) Identify two scales that could be called an "electron charge scale."
   (c) Identify one scale that could be called a "speed of light scale."

Appendix

Answers to Problems

1.1 $\pi_1 = \rho c^3/hv^3$ and $\pi_2 = hv/k_B T$.

1.2 $f = C \cdot v/r$.

1.3 $a = C \cdot v^2/r$.

1.4 (c) $v \propto \sqrt{l}$. (d) For a leg length of 1 meter $v^2/gl = 0.2$.

1.5 $\Delta t = C \cdot l\sqrt{\lambda/\tau}$.

1.6 (a) $\pi_1 = G\rho\Delta t^2$ and $\pi_2 = \rho R^3/m$. (b) $\pi = \rho G\Delta t^2$ and $\Delta t = C/\sqrt{\rho G}$.

2.1 $[G] = M^{-1}L^3T^{-2}$.

2.2 (b) $p = G\rho^2 R^2 \cdot f(r/R)$.

2.3 $\omega* = C \cdot \sqrt{\rho G}$.

2.4 $v = C \cdot \sqrt{MG/R}$.

2.5 $v = C \cdot \sqrt{\tau/\rho}$.

2.6 (a) $D = C \cdot \rho r^2 v^2$. (b) $v = C' \cdot \sqrt{mg/\rho r^2}$. 2.7 $\omega = C \cdot \sqrt{Kl/m}$.

2.9 $R = C \cdot E^{1/5}t^{2/5}/\rho^{1/5}$.

3.2 $\omega = C \cdot \sqrt{\sigma/\rho V}$.

3.3 h/d and $\sigma/(\rho g d^2)$ or, equivalently, $h = d \cdot f(\sigma/\rho g d^2)$ where $f(x)$ is undetermined.

3.4 $t = C \cdot \mu r^2/mg$.

3.5 (a) $\pi_1 = \Delta t(\rho g^2/\mu)^{1/3}$ and $\pi_2 = l\rho g^{1/2}/\mu$. (b) $\Delta t = (\mu/\rho g^2)^{1/3} \cdot f(l\rho g^{1/2}/\mu)$.

3.6 $m = C \cdot \rho v^6/g^3$.

3.7 $\pi_1 = \rho\sigma^3/(\mu^4 g)$, $\pi_2 = v\mu/g$, and $\pi_3 = p\sigma/(\mu^2 g)$ or other complete sets of 3 independent dimensionless products that can be formed from these.

4.1 $[R] = ML^2T^{-2}\Theta^{-1}$.

4.2 (a) $\lambda = C \cdot \sqrt{D_T/\omega}$. (b) $v = C' \cdot \sqrt{\omega D_T}$.

4.3 The dimensionless products are $ql/\kappa\Delta T$ and vlc_p/κ. Therefore, $q = (\kappa\Delta T/l) \cdot f(vlc_p/\kappa)$ where $f(x)$ is undetermined.

5.1 $\omega = C \cdot \sqrt{pE/I}$.

5.2 (a) The dimensionless products are $Ed^2\varepsilon_o/q$ and v/c. Therefore, $E = \left(q/d^2\varepsilon_o\right) \cdot f(v/c)$ where $f(x)$ is undetermined. (b) The dimensionless products are $Bd^2\varepsilon_o c/q$ and v/c. Therefore, $B = \left(q/\left(d^2\varepsilon_o c\right)\right) \cdot g(v/c)$ where $g(x)$ is undetermined.

5.3 $I_{\max}/L = R\varepsilon_o^3 \Delta V^{7/2}/\left(e^{3/2}m^{1/2}\right) \cdot f(e/(s\varepsilon_o\Delta V))$ and $I_{\max}/L = C \cdot \left(\varepsilon_o\Delta V^{3/2}/R\right) \sqrt{e/m}$.

5.5 (a) The dimensionless products are $m\Delta t^2/\left(\mu_o I^2\right)$ and $N_L R$. (b) $\Delta t = \sqrt{\mu_o I^2/m} \cdot f(N_L R)$. In typical pinches $N_L R \gg 1$ and $\Delta t = C \cdot \sqrt{\mu_o I^2/m}$.

5.6 (a)

I	Current	$\left(QT^{-1}\right)$
R	Radius	L
N_L	Density per unit length	L^{-1}
$k_B T$	Boltzmann's constant times absolute temperature	$\left(ML^{-2}T^{-2}\right)L^4$
μ_o	Vacuum permittivity	$\left(ML^{-2}T^{-2}\right)L^3\left(T^2Q^{-2}\right)$

6.1 $E_1 = C \cdot m_e e^4/\left(\varepsilon_o^2 h^2\right)$.

7.4 (a) "Electron mass scales" are based on G, m_e, ε_o, and c, and on G, m_e, ε_o, and h. (b) "Electron charge scales" are based on G, e, ε_o, and c, and on G, e, ε_o, and h. (c) A "speed of light scale" is based on G, ε_o, c, and h. The "speed of light scale" is identical to the Planck scale.

References

1. Galileo Galilei, *Two New Sciences*, translated by Henry Crew and Alfonso de Salvio, (Chicago, IL: Encyclopedia Britannica Inc., 1952), p. 187.

2. H. E. Huntley, *Dimensional Analysis* (Mineola, NY: Dover, 1967), p. 33.

3. Joseph Fourier, *The Analytical Theory of Heat*, translated by Alexander Freeman (Mineola, NY: Dover, 1878), Book II, Section IX, Articles 157–162. Fourier closes his discussion of dimensional analysis with the comment that, "On applying the preceding rule to the different equations and their transformations, it will be found that they are homogeneous with respect to each kind of unit, and that the dimension of every angular or exponential quantity is nothing. If this were not the case some error must have been committed in the analysis . . . ."

4. Lord Rayleigh (John William Strutt), The Principle of Similitude. *Nature*, March 18, (1915).

5. Edgar Buckingham, On Physically Similar Systems; Ilustrations of the Use of Dimensional Equations. *Physical Review*, Vol. IV, no. 4, (1915), 345–376.

6. H. L. Langhaar emphasizes this aspect of the π theorem. See his *Dimensional Analysis and Theory of Models* (Hoboken, NJ: Wiley, 1951), p. 18.

7. Edgar Buckingham, On Physically Similar Systems; Ilustrations of the Use of Dimensional Equations. *Physical Review*, Vol. IV, no. 4, (1915), 345–376.

8. This definition reformulates an equivalent one found in E. R. Van Driest, On Dimensional Analysis and the Presentation of Data in Fluid Flow Problems. *J. Applied Mechanics*, Vol. 13, no. 1, (1946) A–34. See also H. L. Langhaar, *Dimensional Analysis and Theory of Models* (Hoboken, NJ: Wiley, 1951), p. 29.

9. The number of effective dimensions N_D is also the "rank of the dimensional matrix" – a mathematical concept exploited in fluid mechanics engineering texts. See, for instance, R. W. Fox, A. T. McDonald, and P. J. Pritchard, *Fluid Mechanics* (Hoboken, NJ: Wiley, 2004), pp. 282–283.

10. The concept, although not the name, of *imposed dimension* originates with Percy Bridgman's highly recommended text *Dimensional Analysis* (New Haven, CT: Yale University Press, 1922). See, in particular, pp. 9–11, 63–66, 67–69, and 77–78. Others, including H. E. Huntley, *Dimensional Analysis,* (Mineola, NY: Dover, 1967), use the concept of imposed dimensions.

11. G. I. Barenblatt, *Scaling, Self-Similarity, and Intermediate Dynamics* (Cambridge, UK: Cambridge University Press, 1996), pp. 18 ff.

12. Hermann Helmholtz, *On the Sensations of Tone 6th Edition* (Gloucester, MA: Peter Smith, 1948), pp. 43–44.

13. Percy Bridgman, *Dimensional Analysis* (New Haven, CT: Yale University Press, 1922), p. 107.

14. G. Taylor, The Formation of a Blast Wave by a Very Intense Explosion. II. The Atomic Explosion of 1945. *Proceedings of the Royal Society of London. Series A, Mathematical and Physical Sciences*, Vol. 201, No. 1065, (Mar. 22, 1950), 175–186.

15. David C. Lindberg, *The Beginnings of Western Science* (Chicago, IL: University of Chicago Press, 1992), p. 305.

16. See Aristotle, *The Basic Works of Aristotle* editor Richard McKeon, (New York, NY: Random House, 1966) Physics, Book IV, chapter 8, p. 216^a, lines 14–17. See also David C. Lindberg, *The Beginnings of Western Science* (Chicago, IL: University of Chicago Press, 1992), pp. 59–60.

17. R. P. Godwin, The Hydraulic Jump ('Shocks' and Viscous Flow in the Kitchen Sink), *American Journal of Physics* 61 (9), (1993) 829–832.

18. Fridtjof Nansen, *Farthest North* (Edinburgh, UK: Birlinn, 2002), p. 186.

19. C. Vuik, Some Historical Notes on the Stefan Problem, *Nieuw Archief voor Wiskunde*, 4e serie 11 (2), (1993) 157–167.

20. Percy Bridgman, *Dimensional Analysis* (New Haven, CT: Yale University Press, 1922), problem 19, p. 108.

21. Traditionally the "Boussinesq problem." See Percy Bridgman, *Dimensional Analysis* (New Haven, CT: Yale University Press, 1922), pp. 9–11.

22. Lloyd W. Taylor, *Physics: The Pioneer Science* (Mineola, NY: Dover, 1941), pp. 592–593.

23. John Dunmore, *Pacific Explorers: The Life of John Francois de La Perouse 1741–1788* (Palmerston North, NZ: The Dunmore Press Limited, 1985), pp. 286–292.

24. L. Schiff, *Quantum Mechanics 3rd Edition* (New York, NY: McGraw Hill, 1968) pp. 397 ff.

25. Max Planck, *Theory of Heat Radiation* (Mineola, NY: Dover, 2011), pp. 205–206. Number 164.

Index